Effective Business Writing in Ten Minutes a Day

From the Library of:

Alycee R. Nelson Ruley
111 Broad Street
Morton PA 19070

(610) 543-5784

My books are treasured, particularly dear family members. Visit you they may, but please remember, they have a home to return to someday!

Cosmo F. Ferrara, Ed.D.

Effective Business Writing in Ten Minutes a Day

Short, Practical Solutions to
Real-World Writing Problems in
72 Two-Page Lessons

CHILTON BOOK COMPANY
RADNOR, PENNSYLVANIA

Acknowledgments

I would like to thank my wife Fran and my daughter Amy for typing the manuscript for this book and adding their words of humor and encouragement to the effort.

Copyright © 1989 by Cosmo F. Ferrara
All Rights Reserved
Published in Radnor, Pennsylvania 19089,
by Chilton Book Company

No part of this book may be reproduced, transmitted, or stored in any form or by any means, electronic or mechanical, without prior written permission from the publisher

Designed by William E. Lickfield

Manufactured in the United States of America

Library of Congress Cataloging in Publication Data

Ferrara, Cosmo F.
    Effective business writing in ten minutes a day / Cosmo F. Ferrara.
        p. cm.
    Includes index.
    ISBN 0-8019-7979-X
    1. Business writing.   I. Title.
HF5718.3.F47   1989
808.06665—dc20   89-42861
                                                              CIP

1 2 3 4 5 6 7 8 9 0   8 7 6 5 4 3 2 1 0 9

To Mom and Dad

Contents

Listing of Contents by Subject   xi

Introduction   2

1 Hook the Reader   4
2 For Clarity, Write Short Sentences   6
3 Avoid Clichés   8
4 Get the Most from Each Idea   10
5 Make Bulleted Items Parallel   12

   Just Think About It **1**: What Is Good Writing?   14

6 Highlight Ideas with Phrasing Patterns   16
7 Use Comma Sense   18
8 Spelling Does So Count!   20
9 Anchor the Floating "This"   22
10 Repeat Key Ideas, or Once Is Not Enough   24

   Just Think About It **2**: Writing Is a Three-Step Process   26

11 Letters That Remove the Chill from Cold Calls   28
12 Beware of Series of Qualifiers, or Don't String the Reader Along   30
13 Keep the Verb Close to the Subject   32
14 Details, Definitions, and Examples Make Abstract Ideas Concrete   34
15 Omit Unnecessary Words   36

   Just Think About It **3**: Brainstorming in Writing   38

16 Avoid Expressions That Lead to Wordiness 40
17 The Topic Sentence: A Guidepost for Readers 42
18 How to Build an Outline: The Beams and Girders of a Report 44
19 The Hyphen: You've Come a Long Way, Baby 46
20 Transitions: Bridges to Understanding 48

Just Think About It **4**: Breaking the Rules 50

21 The Analogy: Explaining What It Is by Telling What It's Like 52
22 Good Beginnings 54
23 The Middle: Delivering What's Promised 56
24 Writing the Ending, or Leave 'em Laughing 58
25 Keep It Simple: Use Conversational Language 60

Just Think About It **5**: Stop Wasting Time 62

26 Add Vitality with the Active Voice 64
27 Action Verbs Add Snap 66
28 Make References Clear 68
29 *Who*, *Which*, and *That* 70
30 Writing Numbers: Words or Figures? 72

Just Think About It **6**: Writing's Better the Second Time Around 74

31 Persuasion Starts with the Reader 76
32 Positioning Ideas in Sentences 78
33 Imagery Makes Readers See the Point 80
34 You Said It! Using Direct Quotations 82
35 ... And Everything in Its Place: Misplaced Modifiers and Dangling Participles 84

Just Think About It **7**: What's the Point of Punctuation? 86

36 Variety Is the Spice of Sentence Structure 88
37 Distant Relatives: The Colon and the Semicolon 90

38 Unconventional Techniques Perk Up Business Correspondence   92
39 Dashes and Parentheses: The Interruptors   94
40 Avoiding Sexist Language   96

Just Think About It 8: Breaking More Rules   98

41 Building a Working Vocabulary   100
42 Developing Ideas   102
43 Readability and the "Fog Index"   104
44 Reaching Complete Agreement   106
45 Fragments   108

Just Think About It 9: About Form Letters and Boilerplate   110

46 Drawing Comparisons   112
47 Collecting, Selecting, and Ordering Information   114
48 Common Errors in Word Usage   116
49 Jargon: Not for "Outside" Audiences   118
50 Abbreviations: The Long and Short of It   120

Just Think About It 10: The Written Proposal Is a Selling Statement   122

51 Run-Onnnnnnnnnns   124
52 Discriminate and Delete   126
53 More Troublesome Words   128
54 Proofreading Pays Off   130
55 Writing Directions   132

Just Think About It 11: Writing on a Word Processor   134

56 More on Sexist Language   136
57 Setting the Right Tone   138
58 Advice on Adverbs and Adjectives   140
59 How to Interview   142
60 Subordination: Not All Ideas Are Created Equal   144

ix

Just Think About It **12**: Writing the PD (Practice Development) Article 146

Recommended Answers and Revisions for Exercises 148

Index 177

Contents by Subject

FLAVOR

Highlight Ideas with Phrasing Patterns	16
Add Vitality with the Active Voice	64
Imagery Makes Readers See the Point	80
Unconventional Techniques Perk Up Business Correspondence	92
Readability and the "Fog Index"	104

IDEAS

Hook the Reader	4
Letters That Remove the Chill from Cold Calls	28
Details, Definitions, and Examples Make Abstract Ideas Concrete	34
The Analogy: Explaining What It Is by Telling What It's Like	52
Good Beginnings	54
The Middle: Delivering What's Promised	56
Writing the Ending, or Leave 'em Laughing	58
Persuasion Starts with the Reader	76
You Said It! Using Direct Quotations	82
Developing Ideas	102
Collecting, Selecting, and Ordering Information	114
Discriminate and Delete	126
Writing Directions	132
How to Interview	142
Subordination: Not All Ideas Are Created Equal	144

MECHANICS

Use Comma Sense	18
The Hyphen: You've Come a Long Way, Baby	46
Writing Numbers: Words or Figures?	72
Distant Relatives: The Colon and the Semicolon	90
Dashes and Parentheses: The Interruptors	94
Abbreviations: The Long and Short of It	120
Proofreading Pays Off	130

ORGANIZATION

Get the Most from Each Idea	10
Repeat Key Ideas, or Once Is Not Enough	24
The Topic Sentence: A Guidepost for Readers	42
How to Build an Outline: The Beams and Girders of a Report	44
Transitions: Bridges to Understanding	48

SENTENCE STRUCTURE

For Clarity, Write Short Sentences	6
Make Bulleted Items Parallel	12
Anchor the Floating "This"	22
Beware of Series of Qualifiers, or Don't String the Reader Along	30
Keep the Verb Close to the Subject	32
Make References Clear	68
Positioning Ideas in Sentences	78
... And Everything in Its Place: Misplaced Modifiers and Dangling Participles	84
Variety Is the Spice of Sentence Structure	88
Reaching Complete Agreement	106
Fragments	108
Run-Onnnnnnnnns	124

SPELLING

Spelling Does So Count!	20

WORDING

Avoid Clichés	8
Omit Unnecessary Words	36
Avoid Expressions That Lead to Wordiness	40
Keep It Simple: Use Conversational Language	60
Action Verbs Add Snap	66
Who, Which, and *That*	70
Avoiding Sexist Language	96
Building a Working Vocabulary	100
Drawing Comparisons	112
Common Errors in Word Usage	116
Jargon: Not for "Outside" Audiences	118
More Troublesome Words	128
More on Sexist Language	136
Setting the Right Tone	138
Advice on Adverbs and Adjectives	140

WORDING

Effective Business Writing in Ten Minutes a Day

Introduction

When you were preparing for a career in business, you probably had no idea of the amount of writing you would do on the job. To the surprise of many, American business runs on letters, memos, reports, proposals, summaries, briefs, and other kinds of written documents. (And the Fax machine has *increased* the amount of writing we do.) People who studied to work the numbers, arrange the deals, sell the materials, and do so many other kinds of things all come to the same unexpected realization: *They also have to write.*

Effective Business Writing in Ten Minutes a Day was written for those people, the ones who did not major in English but who find themselves spending thirty or forty percent of their time writing to support, explain, or promote their services or products. This book is intended to help people who have little time, who cannot attend a three-day seminar, or wade through a hefty tome on writing. It will work for people who have no interest in winning Pulitzer Prizes but who just want to improve their performance by writing more effectively.

The book focuses on three major characteristics of effective business writing—correctness, directness, and persuasiveness. *Correctness* is important because many people equate sloppy writing with sloppy thinking and dismiss this kind of letter or memo immediately. Writing must be *direct* because people do not have time to plow through pages of prose, no matter how correct it is. And to be effective, writing must be *persuasive*. Whether it be an internal memo announcing a new procedure or a lengthy proposal to a potential client, your writing must convince others to act on your ideas.

If your writing is not as correct, direct, and persusasive as you'd like it to be, this book can probably help you.

The book is short on theory and long on practical tips for the office writer like you. It treats only the most relevant writing problems and covers each in a two-page lesson. For example, there are separate lessons on hooking the reader, using bullet points effectively, saving time when writing, organizing information from many sources, choosing the right word, avoiding sexist language, making ideas clear and forceful, and following and breaking conventional rules of writing. These techniques are illustrated with examples of both good and poor writing drawn from actual business documents covering a wide range of industries and disciplines.

You can do each lesson in about ten minutes—really—during a coffee break, at lunch, or while waiting for a train. The lessons have short exercises so you can apply immediately what you've learned and then compare your work to the suggested answers in the back of the book. Interspersed throughout the book are short essays without exercises that may change the way you think about writing.

Lessons are not grouped by subject and do not follow one another sequentially. Just start with the lesson that's most meaningful to you at the moment, and work your way through the book according to your needs. If, however, you would like to concentrate on a particular aspect of writing, such as Ideas, Organization, Wording, Flavor, Sentence Structure, Mechanics, or Spelling, the contents by subject or the index can quickly point you to all the lessons on that aspect.

It would be great if you could do two or three lessons a week, but even that is not necessary. Each lesson is self-contained, so continuity is not required. If weeks go by between lessons, you'll still be productive without going back over what you've already done. Or, faced with a particular writing problem, just consult the index for the appropriate lesson.

The book was written with you, the real business writer, in mind. It addresses the real needs of people on the job who want to enhance their success with more effective writing. And it recognizes the limits on your time and patience. So, then, let's begin.

IDEAS

1. Hook the Reader

How do you hook readers, generate their interest, and keep them reading? The best way is to focus on their concerns. By nature, we're all selfish, primarily interested in our own welfare, jobs, families. Appeal to our vanity, sense of security, desire for success, love, happiness, and you've got us in the palm of your prose.

Note how the opening to a sales letter addresses the reader's needs:

> One of the problems of living in the Information Age is that there's just too much information. You know the information you want must be stored neatly in some database, but the question is where? You spend so much time tracking down the facts, you have little time for your clients and your work. There is a way, however, to get quality information quickly and easily....

The message goes right to the heart of the reader's concerns. Before saying anything about who's selling what, it hits the reader's "hot buttons"—the need for quality information and to save time. When the reader gets the feeling the writer "knows my business," he or she is hooked and will probably read on.

**To hook the reader,
de-emphasize the "I" and emphasize the "You."**

INSTEAD OF:	TRY
We are happy to announce ...	*You* will be happy to hear ...
We use this approach because ...	*You* benefit from this approach because ...
We serve dozens of companies in your industry ...	Dozens of companies in *your* industry call on us ...

| So that I can process benefits claims more easily ... | So that *your* benefits claims can be processed more quickly ... |

The attitude of most readers is "tell me about my lawn, not about your grass seed." When you want readers to buy your products, services, or ideas, or when you want them to *do* something, *find the slant that focuses on their self-interest.*

Just Try It

Rewrite these sentences to hook the reader through self-interest.

1. We at ABC have developed new products and services to deal with the technological changes assaulting the widget industry.

2. *Office News* needs information on the happenings and hobbies of people in our offi̶ . Help us produce a great newsletter by sending a few paragraphs of your activities.

Write the opening to a memo explaining to a superior why you should be given greater responsibility in the department.

In Your Writing Today...
☞ **Hook your readers; put the spotlight on them.**

SENTENCE STRUCTURE

2. For Clarity, Write Short Sentences

The point of this lesson is so obvious it hardly needs a rule. But the same could be said for "Thou shalt not kill." Unfortunately, more good ideas have been put to a slow death by long, rambling sentences than by any other writing flaw.

Readers should not have to study a sentence to understand it. A sentence that runs *more than three typewritten lines is dangerously long* and a candidate for pruning. All you need to shorten a verbal vagabond is a strategically placed period, or a period and a word change or two. For example:

> TOO LONG: Before any equipment is purchased, uses for it must be determined because unless we know what people want to accomplish with the new equipment, we will not have a solid basis for the purchase or for the design of new systems.
>
> SHORTER AND CLEARER: Before any equipment is purchased, uses for it must be determined. Unless we know what people want to accomplish with the new equipment, we will not have a solid basis for the purchase or for the design of new systems.

To express your thoughts clearly, keep your average sentence length at 18–20 words.

Just Try It

Improve the clarity of the floods of words in these sentences by shortening and damming each with a period and appropriate word changes.

1. The software plan must include a number of recommended applications (such as spreadsheet and other applications) and must address the development of uniform standards so that the information systems department does not have to support multiple technologies for performing the same activity.

2. One popular approach to software selection is to select one or two packages for which training and other user assistance will be provided and let users know that if they want to implement other packages for the application, they do so knowing that they will not be able to call on internal resources for support.

In Your Writing Today...
☞ **Improve clarity. Keep your sentences short.**

3. Avoid Clichés

A cliché is an expression that has become dull through overuse, such as "dead as a doornail," "the bottom line," "the game plan."

At one time these and other clichés were lively, imaginative, effective ways of conveying certain ideas. They were so effective, in fact, that people used them over and over. That's how they became clichés. The problem now is we've overworked these expressions so much, they've lost their polish. The only shine they have is like that on a worn-out pair of pants. Instead of creating a powerful impact on readers, clichés do just the opposite. Readers react with: "If the words are so worn, how fresh can the thoughts be?"

Replacing a cliché may take considerable time and thought, and may require more words than the original. The replacement may be less imaginative and lack the punch of the original. So you have to make a judgment: Do you choose the clear but hackneyed cliché or the fresh but bland replacement? The only way to make this decision is to consider the effect of each on the reader. *It's safer to choose a neutral expression over one that is likely to produce a negative reaction.* "You pays your money and you takes your choice."

But under no circumstances should you use two clichés in the same sentence or paragraph. Readers have little tolerance for sentences like: The company was drowning in red ink until new management came riding in like the cavalry to save the sinking ship.

Just Try It

Rewrite these sentences, replacing the clichés. Your original expressions need not be as picturesque as the clichés, but they should convey the same ideas effectively.

1. Good managers avoid sounding like Monday-morning quarterbacks.

2. The plan seemed workable, but the fly in the ointment was the cost.

3. We can have a ballpark figure for you in a week.

Rewrite this sentence, eliminating one or both clichés.

Though we may still be wet behind the ears in this business, we recognize a pig-in-a-poke offer when we see one.

In Your Writing Today...
☞ **Avoid clichés. If you must use one, use just one.**

ORGANIZATION

4. Get the Most from Each Idea

We often sabotage our own writing. Instead of presenting ideas one at a time, so readers can grasp them, we toss them out in bunches. And instead of explaining each idea fully with the supporting detail it needs, we may string along disconnected one-liners and expect readers to supply what's missing. The result is twofold, and both folds are bad: Readers cannot locate the main ideas, and if they do, they can't really understand them because of a lack of detail.

To get the most out of every main idea, devote at least a full paragraph to each one. State the idea in general terms in a topic sentence; then explain, illustrate, describe, justify, and in other ways clarify the main idea of the topic sentence. For example:

> The company suffered from an inability to control credit. Until 1987 there was no central management of credit. Each of the 900 stores kept its own credit information, and customer payments were made directly to the stores. Terms in many stores were extremely generous. On only the strength of a signature, a customer could buy anything and be given 36 months to pay—at a minimum of $1.00 a month. From 1982 to 1987 accounts receivable ballooned 86 percent to $602 million. Credit losses accounted for 62 percent of the company's deficit.

The writer gets the most out of the main idea—problems caused by the inability to control credit. That idea is hung out in the opening sentence like a ship's flag, so readers can spot it easily. Its full meaning is made clear by the supporting sentences.

Just Try It

The main idea in this paragraph is in the first sentence. Delete information from the other sentences that does not relate directly to the paragraph's main idea.

> Leadership changes aided the collapse. Communication throughout the company was also a problem. The founder's brother-in-law, Harvey Nepotamkin, served as president from 1962 to 1968. These men had never really liked each other but were tied together through the wife/sister, which neither really liked either. Nepotamkin's successor was Mabel Frump. When the expansion program ran into trouble, Frump was replaced by Sidney Stylish at the end of 1971. Inventory problems were quite evident during these years and management should have recognized the credit disaster developing. As sales and profits worsened, company veteran R. J. Fullhouse was summoned from retirement to try to turn things around.

Which ideas in this passage should be main ideas in paragraphs of their own?

In Your Writing Today...
☞ **Present one idea at a time, and explain it fully with supporting detail.**

SENTENCE STRUCTURE

5. Make Bulleted Items Parallel

The bullet-point list is common in business documents to highlight important information. *Be sure, however, the bullets are parallel in construction.* For instance, if the first bullet begins with a verb, the following points should also begin with a verb. For example:

> Our company can help you:
> - *develop* your sales strategies
> - *train* your sales staff
> - *identify* critical issues in the industry.

To make bullets parallel, *rearrange one or more of the elements to match the others in form.* For example:

> Scheduling outside the standard process creates many problems:

> NOT PARALLEL:
> - uneven *distribution* of hours
> - poor *use* of resources
> - *the office may miss its profit plan*

Rewrite the nonparallel element to match the others:

> PARALLEL:
> - uneven *distribution* of hours
> - poor *use* of resources
> - the *possibility* of the office's missing profit plan

When the bullets complete a sentence, each bullet should *fit grammatically* with the lead:

12

It is important that *you:*
- *submit* your schedule on time
- *notify* the schedule administrator of changes
- *request* additional resources as soon as possible

Just Try It

Recast one or more bullets in each sentence below to make all bullets parallel. Also make them grammatically correct.

1. Duties of the department include:

- maintaining files on all clients
- documentation of expenditures
- to report irregularities in office practices

2. The report will provide:

- an analysis of each key issue
- profiles of the market and consumers
- discuss the options open to the company

In Your Writing Today...
☞ **Put all bullet-point items in the same form or parallel construction.**

✔ What Is Good Writing?

Is writing, like wine, a matter of taste? Not so, say judges in a writing experiment. Fifty-four individuals from six occupational fields were asked to evaluate 300 pieces of writing, using their own evaluative criteria. When the lists of criteria were compared, they were found to be almost identical.* Good writing meant the same thing to all fifty-four judges!

The criteria of good writing on every judge's list are:

1. *Ideas.* A well-written paper is rich in ideas. It discusses each point long enough to show clearly what is meant, but no longer. It supports each main point with details that give the reader a reason for believing the idea. No necessary points are overlooked. There is no padding.

2. *Organization.* The paper starts at a good point, has a sense of movement, gets somewhere, and then stops. It has an underlying plan that the reader can follow. Ideas are treated in proportion to their importance, with major points receiving greatest length and emphasis. All points are clearly related to each other and to the main idea.

3. *Wording.* The paper reflects the writer's interest in putting words together to be clear, forceful, and interesting. Words are correct, precise, and imaginative, chosen with an understanding of the readers.

4. *Flavor.* The writing sounds like a person, not a committee or computer. It seems sincere and to be drawn from the writer's own knowledge and experience. It reflects the writer's personality and attitude toward the reader and subject. Some people call this *style.*

5. *Usage and Sentence Structure.* The paper follows accepted forms of usage for written English. Sentence structure is correct

---

*Paul Diederich, *Measuring Growth in English* (National Council of Teachers of English, 1974).

and varied. Sentences are short and simple rather than long and complicated.

6. *Mechanics.* The paper follows the rules of punctuation, capitalization, abbreviations, and numbers. Governing mechanics is the desire to make reading easier and clearer.

7. *Spelling.* The paper is free of spelling errors. The writer takes the time to proofread carefully and consult a dictionary.

8. *Appearance.* The paper is attractive and neat. It makes good use of white space and avoids a crowded look. Readers are drawn to it rather than intimidated by its appearance.

In Your Writing Today...
☞ Think about how to improve in one or more of the eight criteria.

6. Highlight Ideas with Phrasing Patterns

A good interior designer uses lighting or accent pieces to draw attention to a particular table or painting in a room. In much the same way, writers use highly visible word constructions to focus attention on key ideas.

One such construction is *repetition,* in which words or structures are repeated to emphasize a point and link one thought with another. For example:

> Sound business decisions must be made on relevant data. Sound business decisions must be made with a long-term perspective. Sound business decisions must be made on time.

The repetition forces readers to grasp the key point of how business decisions are made. The repetition also links the related concepts (*relevant data, long-term perspective, on time*). Emphasis and linkage are achieved by repeating the word order as well as the words. Naturally, good writers reserve this kind of repetition for the most important ideas.

Another form of verbal spotlight is *antithesis.* This construction expresses *contrasting* ideas in similar arrangements of words, clauses, or sentences. For example:

> I come *to bury* Caesar, not *to praise* him.

> Our success lies not so much *in our products* as *in our vision.*

Each example highlights a contrast through similar phrasing:

to praise Caesar	*to bury* him (infinitive phrases)
in our products	*in our* vision (prepositional phrases)

Just Try It

Use *repetition* in a paragraph to stress the point that *detailed planning* led to the following positive results.

- The building program is ahead of schedule.
- We have controlled costs.
- Construction is expected to be completed within budget.

Use *antithesis* in a sentence to show the difference between two approaches to a job you do.

In Your Writing Today...
☞ **Highlight ideas through repetition and antithesis.**

MECHANICS

7. Use Comma Sense

A famous writer, Anatole France, said he once spent all morning deciding to insert a comma into a sentence. He then spent the whole afternoon taking it out. His productivity for that day was low indeed. To avoid that kind of agony, remember that the comma should help readers separate words and ideas. Insert the comma to tell readers to pause, just as you insert the period to tell them to stop. The comma serves as a pause in two specific instances.

USE A COMMA TO PREVENT MISREADING

Note how the following sentence can easily be misread:

CONFUSING: In deciding to expand the management team broke with tradition.

Readers may easily read "to expand the management team." A strategically placed pause (comma) will prevent the possibility of such confusion.

CLEAR: In deciding to expand, the management team broke with tradition.

USE COMMAS TO ENCLOSE NONESSENTIAL MODIFIERS

A comma is placed *before and after* the nonessential modifier:

NONESSENTIAL: Ms. Weiss, *a partner in our Dallas office*, will handle the case.

The modifier is nonessential because the main idea of the sentence *would not change without it*. If a change would result, the modifier is essential, and no commas are used:

ESSENTIAL: Organizations *that grow too fast* often lose control.

Sometimes the same modifier can be essential or nonessential depending on the intended meaning. In the following examples, the essential modifier *defines* the company; the nonessential modifier merely gives *additional information:*

ESSENTIAL: The company *in Chapter 11* is a client of ours.

NONESSENTIAL: The company, *in Chapter 11,* is a client of ours.

Just Try It

Insert or delete commas to make these sentences clearer.

1. Managers, who straddle the fence, eventually find it uncomfortable.

2. When working the machine produces excellent copies.

3. If all items match the purchase order system schedules payment.

In Your Writing Today...
☞ **Use commas to make writing clear.**

SPELLING

8. Spelling Does So Count!

I read recently of a town whose police officers were required to take instruction in spelling. Why spelling classes for police officers? "Any defense attorney," the chief said, "can get a jury to doubt the testimony of an arresting officer whose report is marred with misspellings." In business correspondence, misspelled words jump off the page. Even some of the worst writers can spot a misspelled word at a glance, and then doubt the competence of the writer in *all* areas of business.

To rid your business documents of misspelled words try these four steps:

PROOFREAD

Most spelling errors result from typos or carelessness (for example, *who's* for whose, to for *too*, *thh* for *the*.) After you have revised the content, form, and wording, proofread the document for spelling errors. *To focus on spelling and not ideas, read from the bottom up.*

USE A DICTIONARY

I know. You're about to say, "How can I look it up if I don't know how to spell it?" You must have a rough idea of the spelling, so check that first. If that's not correct, consider the alternatives. Usually people get confused with double letters, *e* or *i*, *s* or *c*, and such.

KEEP A LIST

Make a list of the words you use often and have trouble spelling. Refer to that list (with correct spellings, of course) as needed.

APPLY THE RULES

Despite the exceptions, rules of spelling can help. For example, you can prevent many errors by applying the best-known rule: Write *i* before *e*, except after *c*, or when sounded like *a*, as in *eighty* and *sleigh*.

> i before e: bel<u>ie</u>ve, w<u>ie</u>ld
> ei after c: rec<u>ei</u>ve, c<u>ei</u>ling
> ei when sounded like "a": w<u>ei</u>gh, fr<u>ei</u>ght
>
> Some Exceptions: finan<u>cie</u>r, l<u>ei</u>sure, effi<u>cie</u>nt, w<u>ei</u>rd

Just Try It

Proofread this paragraph for *ten* misspelled words. Underline them and write the correct spellings below.

> The inventory problem was intensifyed by inefficeint managment. Inadequatly-trained managers were oparating thier stores almost independantly. Buyers had to rely on inacurate sales figures, which resultd in too many under- or over-stocked conditions.

In Your Writing Today...
☞ **Spell every word correctly.**

SENTENCE STRUCTURE

9. Anchor the Floating "This"

The word *this* is a major source of confusion. *This* normally is used with another word: this problem, this candidate, this approach. *This* may be used alone, but only if its referent (the word it refers to) is absolutely clear. For example: "The temperature in the computer room was 85 degrees. *This* was entirely too high." The context and the close proximity of *this* to 85 degrees leave no doubt about the referent.

Confusion reigns if the reader must ask "this what?" For example:

> CONFUSING: The new marketing plan calls for the addition of a telemarketing unit. This has been developed after careful research and analysis.

This what? This unit? This addition? This plan? Eliminate ambiguity by anchoring the floating *this* to its referent:

> CLEAR: The new marketing plan calls for the addition of a telemarketing unit. *This plan* has been developed after careful research and analysis.

Confusion arises when: *this* is too far removed from its referent; or two words or ideas might be perceived as the referent; or there is no suitable referent within miles.

Sometimes you must supply a *word* to go with *this* because the referent is not a single word but an idea.

> CONFUSING: The policy on assignment of office space has been revised. This should prevent the bickering we have experienced recently.

This *what*? This policy? This assignment? Actually *this* refers to the *idea* of policy revision. It is the revision that will prevent disruption. So the passage should be written:

CLEAR: The policy on assignment of office space has been revised. *This revision* should prevent the bickering we have experienced recently.

Just Try It

Anchor the floating *this* in these sentences.

1. We have devised a unique approach to deal with the issue of employee retention. This is best described in stages.

2. In recent months the personnel department has emphasized attitude over skills in hiring entry-level staff. This has already shown some positive effects.

In Your Writing Today . . .
☞ **Ask "this what?" Then anchor the floating *this*.**

10. Repeat Key Ideas, or Once Is Not Enough

Have you ever been alone in a car with the radio on and not had the slightest idea what you've been listening to? We all have this tendency to tune out. That's why copy writers repeat the names of products, sale dates, and phone numbers in commercials.

Business writers would do well to follow the same practice. Surprising as it may sound, readers do not hang on our every word. They may *read* every word but still overlook key ideas or fail to grasp their meaning.

To be sure your readers seize every important point, repeat key thoughts in key paragraphs.

For example:

A chief cause of employee turnover is their lack of commitment. For employees to gain that commitment, supervisors must define goals for employees and stress their importance in the company mission. Employees will feel more a part of the company and see purpose in their jobs. Without this feeling of self-worth and a sense of purpose in the operation, employees are more likely to go looking elsewhere to find them.

The main idea about commitment and turnover is presented in the first sentence and repeated in the last. The wording is changed, as it should be, but the idea is the same.

This kind of repetition (call it "paraphrasing") does not belabor a point or write "down" to readers. But it does help the reader and ensures full comprehension.

Just Try It

Write a closing sentence for this paragraph that repeats (paraphrases) its main idea.

> Despite the costs, turnover does at times have positive aspects. For example, turnover provides the opportunity to replace below-average performers with above-average performers. When such a switch is made, productivity can increase greatly, more than making up for the costs of terminating, recruiting, and training.

In Your Writing Today...
☞ **Repeat key ideas in key paragraphs.**

✔ Writing Is a Three-Step Process

Writing is best done in three steps: generating, drafting, and revising. In the first two steps you are trying to understand the topic and what you want to say about it. In the third step you search for ways to communicate it to your audience.

In the *generating* stage you are actually writing, though not for the reader. You are jotting notes, scribbling figures, drawing diagrams, trying out sentences and phrases. The generating stage of writing helps you put random thoughts on paper so as not to lose them. It also primes the pump, gets ideas flowing, and increases your understanding of the topic. You may also brainstorm (with others or alone), do research, or go for a walk to sort out your thoughts. You gather data and ideas from reading, discussions, and thinking; then you begin to interpret findings and tie bits of information together.

Throughout this generating step you are not concerned about explaining the topic to the reader but rather learning more about the topic yourself. Only when you feel you have a good grasp of the topic do you start to organize the material. First, sort out notes and scribbles according to themes or subtopics. From there you can create a formal outline.

Now you are ready for *drafting*, or taking a first step at putting ideas into a unified, flowing prose form. Work from your notes (and the outline), turning the bits and pieces you collected earlier into complete sentences. You may add new thoughts as they occur and decide that some points no longer fit in the scheme that is developing. You can make those decisions now because the act of drafting takes you to a higher level of understanding than generating did.

In the drafting stage you are still writing primarily for yourself, to determine precisely what you have to say and what you *want* to say. That's why you are not too concerned in your draft with wording, sentence structure, and other fine points. You are concentrating on *substance*.

The third step—*revising*—moves the writing from *writer*-centered to *reader*-centered. Until now, the writing process has helped *you* understand the material; now you must fashion it for the reader. If possible, leave the draft for a while and then try to view it through the eyes of your reader. Anticipate the reader's needs, reactions, and questions. Make whatever changes are necessary to clarify your thoughts for the reader. Those changes may be major, such as reordering, adding, or deleting large segments of information. Or they may be less severe, such as inserting a transition, breaking a long sentence into two, or correcting errors in grammar, usage, word choice, and spelling.

Knowing that the process involves three steps, you will not try to do too much too soon, and you will not panic if after two hours of work you do not have two pages of perfect prose. Employing the three steps in the writing process helps you search for the best substance, structure, and style to meet the demands of purpose, audience, and context.

In Your Writing Today...
Employ the three steps in the writing process.

11. Letters That Remove the Chill from Cold Calls

If you are involved in the marketing or selling end of your business, you have suffered the chill of the cold call. Approaching potential customers sight unseen, over the phone or in person, can be disheartening. A good sales letter, however, can open the door and establish a warmer welcome.

The sales letter attempts to:

- introduce you and your product or service to the prospects
- make them curious about you or your product or service
- pave the way for a personal meeting

CREATE A MILD SENSE OF URGENCY

Creating a mild sense of urgency will get prospects to read your sales letter. Begin your letter with concerns and problems in the prospect's industry just to pique the reader's interest. Focus on a few issues the company is probably facing. Present these issues in a short paragraph or two.

IDENTIFY BENEFITS

Next, indicate briefly how you or your service can help the prospect address these issues. Don't tout your company, but speak of its offerings in terms of the reader's needs.

SCHEDULE A CALL

To close, suggest that the reader would find a meeting with you helpful. Don't try to sell now, and don't invite the reader to call you. State when you will call to arrange this meeting, preparing the prospect and increasing your chances of being received.

KEEP IT SHORT

Make your letter inviting to read, not intimidating. Limit it to *one page*, with plenty of white space. Use bullet points, underlining, and boldface to call attention to key ideas.

Just Try It

Rewrite the beginning of this sales letter.

> As pioneers in the widget industry, we at WOW Widgets have always tried to meet the needs of our customers. We continue this practice with a new service line for inventory control called WOW-I.

In Your Writing Today...
☞ **Focus sales letters on the customer's needs.**

12. Beware of Series of Qualifiers, or Don't String the Reader Along

Business writers often cause confusion and force rereading when they hold off a key noun, tantalizing readers with a series of adjectives or qualifying words. For example:

> CONFUSING: These changes strengthen our commitment to a highly participative, field-driven decision-making organization.

The object of *to* is *organization*. But with so many words between them, readers may not realize that.

> CLEAR: These changes strengthen our commitment to a decision-making organization that is highly participative and field-driven.

Confusion is more likely when the qualifers are nouns used as adjectives, because these are perceived as the key words. For example:

> CONFUSING: Phase II calls for *systems requirements* definition.

Readers may misread the sentence, assuming Phase II calls for systems, or systems requirements. To prevent even the possibility of confusion, move the key word up front:

> CLEAR: Phase II calls for *definition* of systems requirements.

Here's another example, this time with a committee name:

CONFUSING: The Financial Accounting Standards Board's Emerging Issues Task Force recently reached consensus on stock purchase plans.

The key word here, the actual doer, is the *Task Force,* or to be precise, the Emerging Issues Task Force. Readers would not know that, however, and would probably assume the Financial Accounting Standards Board is the doer. To prevent that kind of confusion, put the doer up front.

CLEAR: The Emerging Issues *Task Force* of the Financial Accounting Standards Board recently reached consensus on stock purchase plans.

Just Try It

Rewrite these sentences, putting the qualifiers after the key word.

1. Segregating duties will help ensure the proper safeguarding of cash receipts and insurance premium collections.

2. A committee has been formed to examine long-term strategy and short-term objectives costs and benefits.

In Your Writing Today...
☞ **Put the key word before the string of qualifiers.**

13. Keep the Verb Close to the Subject

One reason Americans find the German language difficult is that verbs are often put at the end of a sentence. For example, the sentence "We have taken a physical inventory" would be written, "We have a physical inventory taken." Holding in abeyance the key word "taken" is awkward if you're not used to it.

Business writers often create a similar confusion when they separate the verb from its subject. By the time readers get to the verb, they have forgotten the subject. For example:

> CONFUSING: The most critical *factors* in a business plan's projections, if the business plan is being used to obtain financing rather than for charting the course of the business, *are* the assumptions behind them.

Putting the verb closer to the subject makes the sentence clearer:

> CLEAR: If the business plan is being used for financing rather than for charting the course of the business, the most critical *factors* in a business plan's projections *are* the assumptions behind them.

Just moving the verb and what follows next to the subject may not always be possible. You may have to *restructure the sentence and change some words* to get a tighter, clearer sentence. For example:

> CONFUSING: A *plan* that provides for the study and evaluation of the department at present and an assessment of the strengths of each staff member *should give* the new director an understanding of the task before her.

> CLEAR: The new *director should gain* an understanding of the task before her from a plan that provides for the study and

evaluation of the department at present and an assessment of the strengths of each staff member.

Just Try It

These sentences suffer from subject-verb separation. Rewrite them to make them tighter and clearer. Restructure the sentence if necessary.

1. A review of relevant statistics, such as orders placed, percentage of out-of-stocks, numbers of damaged shipments, frequency and length of delays, and level of bad-debt experience with each supplier, helped us assess our supplier relationships.

2. One common practice that companies opening new branches engage in to ease the burden of large pre-opening costs is to amortize pre-opening expenses over a period of six months to two years.

In Your Writing Today...
☞ **Keep the verb close to the subject.**

14. Details, Definitions, and Examples Make Abstract Ideas Concrete

Good ideas are often ignored or misunderstood because they are left in the abstract. Readers cannot perceive their full meaning unless the ideas are put into concrete situations with *details* and *examples,* and unless *terms are defined.* Note, for example, how a key idea in the following passage is glossed over, leaving readers with a major question.

> CONFUSING: In 1986, management felt the company was ready for expansion. So it laid out a plan to open forty-eight new units over eighteen months.

The company's "readiness for expansion" is a key abstract idea, one that should be explained in more detail. Readers will then understand what "ready" meant to management.

> CLEAR: In 1986, management felt the company was ready for expansion. *Profits had been consistently good, all systems were* **state-of-the-art** *and expandable, and a corps of experienced but young* **managers was** *eager to grow further with the company.* So management laid out a plan to open 48 new units over 18 months.

Another way to clarify ideas is to define your terms, to be sure readers know who or what a term refers to. For example, in the following sentence, note how helpful the definition of the term "end-users" is:

> Accounting policies are important because they form the framework under which financial statements are interpreted by end-

users—*banks, investors, company management, industry analysts, and regulatory agencies.*

Like details and definitions, examples help readers put abstract ideas in a concrete context. Note in the following passage how the example gives a richer, more exact meaning to the abstract idea.

Accounting policies are important because they form the framework under which financial statements are interpreted by end-users—*banks, investors, company management, industry analysts, and regulatory agencies. For example, management seeking capital must present its business in a way potential investors can understand.*

Just Try It

Rewrite this passage, clarifying abstract ideas with details, definitions, and examples.

> The widget industry is becoming more competitive. And recent mergers and acquisitions have created a protectionist mentality among management. In the process, established companies have taken to some very questionable practices.

In Your Writing Today...
☞ **Clarify ideas with definitions, details, and examples.**

35

15. Omit Unnecessary Words

One reason Charles Dickens's books are so long is that he wrote them as serials for magazines and was paid by the word. While it was profitable for him to be wordy, business writers can lose readers—and *money*—because of excessive length. It pays, then, to keep things short and crisp by omitting unnecessary words.

One major source of unnecessary words is *redundancy*. For example:

> WORDY: Morale has reached a serious low point and *employees are not happy in their work.*

Does the second part of the sentence add anything to the first? Not really, so leave it out. Repeating an idea for emphasis is helpful; repeating it out of carelessness is not.

Check your sentences for words you don't need

> WORDY: Solutions *designed to correct* the problem have come from various quarters.

"Designed to correct" is inherent in the meaning of "solution." Therefore those words are unnecessary:

> CRISP: Solutions *to* the problem have come from various quarters.

Here's another example:

> WORDY: Measure the accuracy of the estimates, particularly *for those in which* equipment costs are involved.

There is a shorter, clearer way of expressing this idea:

> CRISP: Measure the accuracy of the estimates, particularly *when* equipment costs are involved.

Just Try It

Omit the unnecessary words from these sentences.

1. Their annual contribution has been $1,000 a year.
2. They developed a comprehensive job description listing all duties connected with the position.
3. Their forecasts of future sales trends are overly optimistic.
4. The program helps managers prepare the plan and monitor the success of the plan.

In Your Writing Today...
☞ **Omit all unnecessary words.**

✔ Brainstorming in Writing

Getting started is perhaps the most difficult aspect of writing. When your information is skimpy and you are not sure what you need, or when you have no idea of what direction to take, try brainstorming. As in a brainstorming discussion with colleagues, the purpose of brainstorming in writing is to generate ideas *in quantity*. Don't judge your ideas at this point. Simply try to write down as many ideas as possible so you can look at them and consider their potential.

One way of brainstorming for writing is through *clustering*. Begin by writing the task or subject in a circle in the center of a blank page. Around it write words or phrases that have even a remote connection to the subject. Expand the cluster as one thought triggers another. Do not try to group, organize, or evaluate ideas until you have exhausted your reservoir of thoughts. Then look for relationships and connect related thoughts with lines. Or you can color-code related thoughts with markers, or enclose them with the same kind of design. (Do not color-code *and* use designs unless you have a very liberal-minded boss.) What you have now is the beginnings for an outline and your first draft.

A brainstorming cluster for a report proposing the "Implementation of a New Procedure" might look like this:

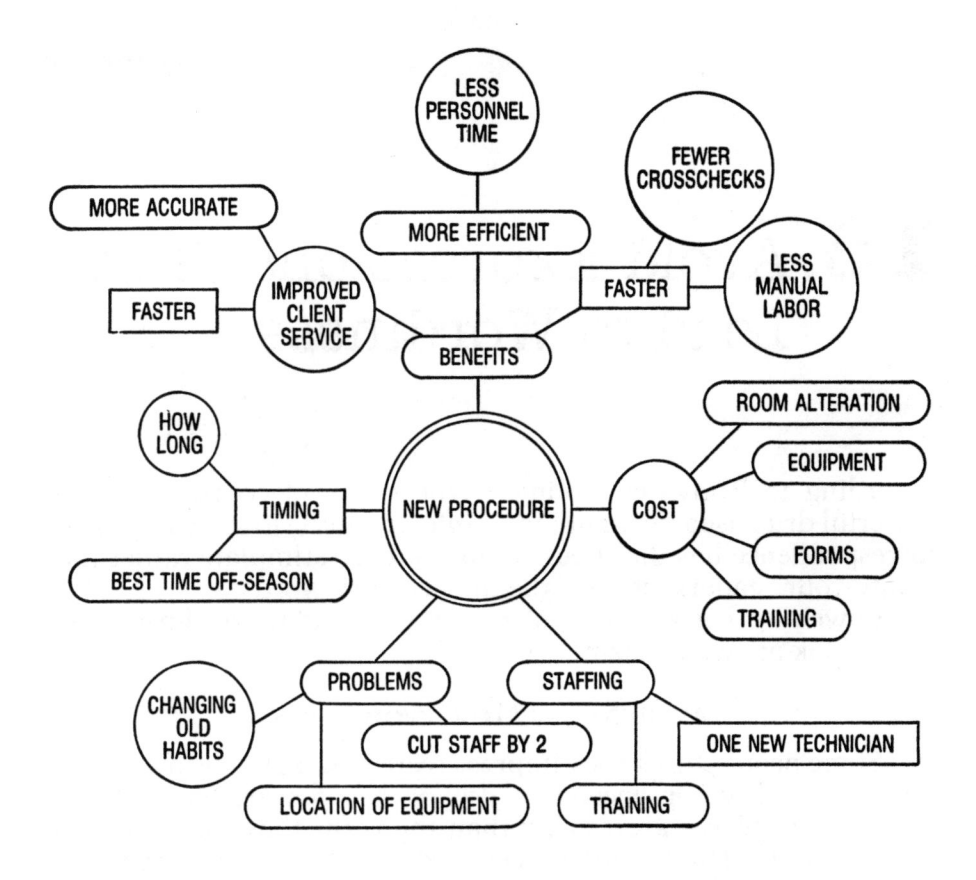

```
                        LESS
                     PERSONNEL
                        TIME
                                      FEWER
                                    CROSSCHECKS
  MORE ACCURATE      MORE EFFICIENT
                                              LESS
              IMPROVED         FASTER        MANUAL
   FASTER      CLIENT                         LABOR
              SERVICE
                        BENEFITS

                                         ROOM ALTERATION
      HOW
     LONG                                    EQUIPMENT
              TIMING    NEW PROCEDURE   COST
                                              FORMS
        BEST TIME OFF-SEASON
                                             TRAINING

              PROBLEMS      STAFFING
    CHANGING
      OLD
     HABITS        CUT STAFF BY 2      ONE NEW TECHNICIAN

           LOCATION OF EQUIPMENT    TRAINING
```

This form of brainstorming may not look like writing, but it is. It involves putting pen to paper, and that step alone has much to do with generating ideas. You are then ready to work those ideas into a more formal outline.

In Your Writing Today...
☞ **Brainstorm to get started.**

16. Avoid Expressions That Lead to Wordiness

According to Rudyard Kipling, "Words are, of course, the most powerful drug used by mankind." Unfortunately, too much business correspondence has the effect of Sominex. To stimulate rather than sedate your readers, cut out the unnecessary words.

In writing, certain kinds of expressions lead to wordiness. Let's take a look at two of them.

Avoid Nouns Made From Verbs

Some people think they impress readers with nouns made from verbs, words like *implementation, development, formulation*. But these "five-dollar" words are weaker than the verbs they sprang from—*implement, develop, formulate*. They are also longer and need other words to shore them up. For instance:

> WORDY: The changes will *lead to the enhancement* of the practice.
>
> CRISP: The changes will *enhance* the practice.
>
> WORDY: We *conducted an evaluation* of the entire process.
>
> CRISP: We *evaluated* the entire process.

Consider *-ment, -ance,* and *-tion* as red flags that signal wordiness.

Avoid Lengthy Phrases

Sometimes we use phrases that are longer than needed.

FOR INSTANCE: INSTEAD OF:
    in order to   to
    for the purpose of   to
    in light of the fact that   because
    in the event that   if
    afford me the opportunity to   allow me
    I am of the opinion   I think

Just Try It

1. What verbs can replace these wordy noun/verb expressions?

    make provisions for   _____

    provide documentation   _____

    undertake completion   _____

2. Provide a word or shorter phrase for each of the following:

    subsequent to   _____

    due to the fact that   _____

    with the exception of   _____

In Your Writing Today ...
☞ **Avoid expressions that lead to wordiness.**

17. The Topic Sentence: A Guidepost for Readers

When shopping in a supermarket, you've probably made use of the aisle directionals, those overhead signs listing the kinds of products in each aisle. With so many items on the shelves, shoppers need that kind of directional help to find the dog food or paper towels without wandering all over the store. Good writing gives similar directionals to help readers navigate through a sea of ideas and information. In writing, a primary directional is the topic sentence.

The topic sentence is just that—a sentence that identifies the topic, or main idea, of a paragraph. The topic sentence lets readers know what the paragraph is about. The reader then has little trouble seeing the significance in a fact or the point of an example in the paragraph. Note how the topic sentence in the following paragraph pulls isolated facts together and gives them a *common* meaning.

> Two years ago the grounds surrounding the plant were strewn with years-old debris. Today those grounds are clean. Noise levels from the factories' machines have been reduced by more than fifty percent. Smoke converters have been installed throughout the building, and our neighbors can once again hang their wash out to dry. Our public relations department meets monthly with municipal government and EPA officials. *The company is fully committed to improving the environment in and around the plant.*

In this example the topic sentence comes last. Often it comes first, making it even easier for readers to follow the thought from the outset.

If the topic sentence contains only one idea, you will not wander off to write about anything else. If the sentence is written in general terms, it gives the *main* idea, and it points toward specific supporting details you should include. Note these examples of faulty topic sentences:

TOO MANY IDEAS: Your organization requires an approach that is responsive to sudden changes in the economy and conducive to cross-fertilization among the staff.
One may be ignored

NOT GENERAL: Sales have increased ten percent over the last year, but expenses have increased by twenty percent.
What's the main idea?

Just Try It

Tell what's wrong with these topic sentences.

1. We added seventeen people to the office staff in the last year.

2. Our company has a bottom-up approach to planning and a zero-based budgeting philosophy.

Rewrite the second sentence above as a topic sentence and build a paragraph around it.

In Your Writing Today ...
☞ **Guide readers with topic sentences.**

18. How to Build an Outline: The Beams and Girders of a Report

There's an apocryphal story of the salesman who was called in for a tax audit. He tossed all his canceled checks, receipts, and other records into a paper bag, hoping the jumble would discourage the IRS auditor from delving into his deductions. When she saw the brown paper jumble, however, the auditor said, "You can sit here and organize your papers. Let my secretary know when you're finished."

The IRS may be undaunted by disorganized data, but most business readers are not. One means of organizing a long document is to outline it. The outline lists in an orderly progression the major elements in the report. Like the beams and girders of a building, an outline gives an overall structure, which you can fill in and complete later. The outline provides a visual presentation of ideas and information planned for the paper. In this form you can spot points out of sequence, arguments that lack support, gaps that must be filled.

How to Outline

Begin to outline *after* you've collected most of your information. If you try to outline too soon, you may ignore key areas of information. Sort that information by category or topic. In describing a business, for example, your categories might be: *personnel, objectives, performance*. The categories become the major points in the outline.

Next, go through the same sorting process *within* each category. *Personnel* might be subdivided into: *needs, current levels, turnover statistics*. These subdivisions identify the supporting matter that explains the main point.

Then develop a *preliminary* outline. (As you write you may decide to modify it.) This conventional lettering and indenting scheme enables you to show relationships:

Title of Report—*An Assessment of the Company*

Roman Numeral	I. Company History	I. major point/topic head
Upper-Case Letter	A. Early Formation	A. categories of supporting detail
	B. Expansion	
Arabic numeral	1. territorial	1. specific supporting details
	2. products and services	
lower-case letter	a) consumers	a) subdivisions of supporting details
	b) businesses	

Just Try It

Complete the outline started above for "An Assessment of the Company." Consider the additional major points to be "current status" and "plans for immediate future."

In Your Writing Today ...
☞ **Organize long reports by outlining.**

MECHANICS

19. The Hyphen: You've Come a Long Way, Baby

Until the 1970s, the hyphen was an insignificant punctuation mark. Then women started tacking their husbands' names to their own, using the hyphen to form combinations such as Gretchen Dopplemayer-Jones and Mary Smith-Bacciagalupe.

Prior to the hyphen's being enlisted in the fight for equality, its primary function had been to mark the separation of a word at the end of a line:

> We thought it best that the local offices be responsible for distribution of the marketing brochures.

To divide words at the end of a line, *break the word between syllables*, as in *in-gen-u-ity*, or *ef-fec-tive*. Do not split a word before *-ed* or *-ly*. When in doubt about syllabification, check the dictionary.

THE HYPHENS IN COMPOUNDS

Some other common uses of the hyphen are in compounds with:

1. Consecutive vowels, as in *re-institute, co-opted*

2. Numbers, as in *thirty-four, one hundred twenty-eight*

3. *Well,* when it and the adjective cannot be reversed, as in *well-disposed* or *well-founded*

4. An adjective and a noun to which *-d* or *-ed* has been added, as in *open-minded*

Some compounds take hyphens when they modify a noun that follows:

- a number and a noun, as in *two-day* conference

- *well* and a participle, as in *well-prepared* proposal
- adjective and participle, as in *far-reaching* effects

When these compounds do not precede the noun, do not use a hyphen:

Effects are *far reaching.*   The proposal was *well prepared.*

Some longer compounds used as adjectives preceding the nouns they modify do *not* need a hyphen:

>*business reply card*   *credit card privileges*
>*direct mail selling*

Words with prefixes and suffixes are usually written as one word:

>*overdue*   *recount*   *businesslike*
>*semiskilled*

There are more rules on hyphens than these and twice as many exceptions. The trend today, however, is toward *less* rather than more hyphenation. When in doubt, check your dictionary.

Just Try It

Insert, delete, or reposition hyphens as needed in these sentences.

1. The company is now planning to reinstate the eight cylinder engine.

2. Twenty four staff people attended the three day meeting that was intended to unveil the well-kept secret of the home-shopping subsidiary.

3. In this labor-intensive environment, lay-offs are common.

In Your Writing Today ...
☞ **Use hyphens correctly.**

ORGANIZATION

20. Transitions: Bridges to Understanding

When a country or a company elects a new president, a major concern is a smooth transition from one administrative team to its successor. Good writers have a similar concern for the smooth flow of ideas from sentence to sentence and especially from paragraph to paragraph. *Connectives* and *repetition* are the primary conduits for facilitating the flow and progression of thought.

CONNECTIVES

The simplest transitions are often the most effective. Connectives like *in addition, however, in the meantime, on the other hand, then, next, first, second,* or *finally* act as a bridge between sentences or paragraphs. These kinds of transitions signal the flow of thought. For instance, they alert the reader to:

- a continuation of a line of thinking (*in addition, furthermore*)
- a shift in thinking (*on the other hand, however*)
- a result or conclusion (*therefore, consequently*)
- a forward movement in a sequence (*next, second*)

One of the most vital connectives is *for example*.

TRANSITIONS THROUGH REPETITION

Another means of helping your readers move from one thought to the next is to *repeat a key word*. For example:

The move to our new headquarters will no doubt *increase* commuting costs for many of our staff. This *increase* will be dealt with on a case-by-case basis.

Sometimes the transition can be achieved by *repeating the idea but in different words*. For example:

> Because of the decline in sales, all departments must explore all possibilities for *cutting expenses*. A company-wide *belt-tightening* will require adjustments, but it is a preferred alternative to reducing staff.

Belt-tightening picks up the idea of *cutting expenses*, but in different words.

Just Try It

1. Use a transitional word or phrase to link these two sentences.

   Office romances have always been thought to affect productivity adversely. Some psychologists claim people in love with co-workers have better morale in the office and better attendance records.

2. Repeat a key word or idea to link these two sentences.

   Though the content of the brochure is acceptable to the committee, there are some serious disagreements about the format. A meeting will be scheduled for Monday at nine.

In Your Writing Today ...
☞ **Use transitions to link sentences and paragraphs.**

✔ Breaking the Rules

Was it George M. Cohan who said there's a broken rule of composition for every light on Broadway? Rules actually codify good writing practices and *are* worth following. Sometimes, however, breaking a rule produces clearer, more effective writing. The key is knowing when to break the rule and why.

NEVER END A SENTENCE WITH A PREPOSITION

This rule is rooted in sound logic. Readers stop on the last word in a sentence, so it's this one they will most likely grasp and remember. Why waste such a vital position with a do-nothing word like a preposition—*with, of, from?* Fill that position with a power word such as a noun—*profit, money*—or a verb—*drive, sell.* At times, however, holding to this rule creates an unnatural, awkward expression: "This is a situation up with which I will not put." The sentence is smoother and more natural with the preposition at the end: "This is a situation I will not put up with."

NEVER BEGIN A SENTENCE WITH *And* OR *But*

This rule follows the same logic as Rule 1. Never waste the *first* position in the sentence. Conjunctions such as *and* and *but* usually do little more than join one word, phrase, or sentence to another and don't command a prominent place. But sometimes the context calls for *a critical addition or a significant change in direction*, which the writer can emphasize by putting the conjunction first.

NEVER USE *I* IN A BUSINESS DOCUMENT

This rule has less validity than the first two and reflects a formality that left American business years ago, at least since we began wearing blue shirts to the office. It is absurd to refer to yourself in a business letter as "this writer" or some equally silly paraphrase. (Doing so conjures up memories of that pompous sportscaster: "In the eyes of this reporter....") *I* is clearer, and more natural and personal.

Never Address the Reader as *You*

If you blindly follow this rule, you are apt to stumble into some stilted prose. If the document is intended for a specific audience, don't try to hide that fact. Make the communication more personal and natural—and more readable—by referring to the reader as *you*. In giving directions especially, it makes more sense to say, "Next, you poll the people in the office..."—not "Next, one polls the people in the office..."

In Your Writing Today ...
☞ **Break the rules when it helps your writing.**

IDEAS

21. The Analogy: Explaining What It Is by Telling What It's Like

One way to simplify complex ideas and enrich common ones is through *analogy*. An analogy is a comparison that allows you to explain one idea by means of another, usually a more familiar one. Though the two ideas have many differences, they are similar in at least one key area. When readers recognize that point of similarity, they grasp the new idea or see an old idea in a new, dramatic light.

Sometimes the analogy appears in a single sentence:

> Why not let our staff, which must live with the solution, work on the problem, instead of bringing in an outside consultant, a hired gun?

The attributes of the hired gun, who is known to have loyalty to no one but himself, are being applied to the consultant, who may be less known to the reader. By using the analogy of the hired gun, the writer explains her position on the loyalty of outside consultants more emphatically than if she had spoken of loyalty in the abstract.

Note how the writer uses analogy in the following passage to explain her opposition to a corporate move:

> Relocating the corporate headquarters from the city to the suburban campus will be like taking a vacation in the country. At first the quiet, the fresh air, and the green surroundings soothe the frazzled nerves. The pace slows and the body begins to unwind. The mere fact of the distance between you and all you have left behind is an added comfort. But then after a few days, the surroundings begin to take on a sameness; the quiet deepens

and deepens. The slow pace has ground to a standstill and you're following the handyman around looking for things to do. And worst of all, you feel so far away, so far from the things you feel you should be doing. The suburban campus may look nice now, but after six months, will we be clamoring to get back to where the action is?

In this passage, the writer relies on readers' knowledge of the vacation syndrome to explain the problems of the suburban relocation.

Make ideas tangible through analogies.

Just Try It

1. Write a single sentence that uses analogy to explain the idea that a direct approach is often the best approach.

2. Write a paragraph built around an analogy. The main idea to be developed in the paragraph is your concept of a good manager.

In Your Writing Today ...
Use analogy to explain and strengthen your ideas.

22. Good Beginnings

If you don't get readers' attention at the beginning, they probably won't read any further. To make sure readers get to the middle of the document, your beginning should accomplish at least one of four things:

IDENTIFY THE SUBJECT AND PURPOSE

A good beginning lets readers know the subject, issue, or problem and your purpose in addressing it. A strong beginning tells readers what they are about to read and why they should read it.

The Bureau has investigated the charges you made in your letter of July 10 and would like to respond to them.

IDENTIFY THE WRITER

A good beginning lets readers know who the writer is, what his or her involvement is with the issue, and what authority he or she has to make the statements that follow. Readers will know they are reading a document of considerable merit.

Here is my response to the proposal for expanding the department. Before making my evaluation I spoke to all the parties involved, examined pertinent records, and considered the pros and cons addressed in the proposal.

TELL WHAT THE DOCUMENT COVERS

A good beginning lets readers know what areas are covered and roughly where they can be found in the document.

As you requested I have prepared an estimated budget for the Department for the next fiscal year. This document includes a summary of expenses by work area, followed by an item-by-item breakdown of these expenses.

Generate Reader Interest

A good beginning generates reader interest. If the subject cannot do that in and of itself, write a beginning that focuses on the readers and their concerns.

These days of high costs demand high efficiency in every aspect of a ship's operation. Economy, however, is difficult with older ships such as your *Pequod II*.

Just Try It

Write the beginning of a memo from the office manager to the staff announcing a change in an administrative procedure. Indicate which of the four objectives of a good beginning you have reached.

In Your Writing Today ...
☞ **Write beginnings that stimulate reading.**

23. The Middle: Delivering What's Promised

Americans pay too little attention to their middles—in writing as well as eating. In a written document the middle should present in detail the subject introduced in the beginning.

PROMISE AND DELIVERY

Too many middles deliver little of what was promised in the beginning. This can happen when what seems like a minor point in the outline absorbs more attention than the writer had intended. For example, in a paper "How to Forecast Cash Flow," the main purpose is to teach readers how to calculate such forecasts. "Reasons for forecasting cash flow" is listed as a preliminary but minor point in the middle. But the writer may get so involved in the many reasons why a company should know how to do this forecast that he gives only passing reference to how it is done. The reader then feels confused, if not cheated. *How-to* was promised, but *why* was delivered.

BALANCE AND PROPORTION

To deliver what you promise, pay attention to balance and proportion. Allocate space by degree of importance. If your purpose is "How to Calculate," give most space to the "how-to," not "reasons for."

ORDER

Order is another consideration for your middle. Will you proceed *logically*, from basic to advanced, for example? Or *chronologically*, from first to last? You might decide on an *order of importance*, working from most to least important, or vice versa. *Order of recognition* makes sense when readers are not familiar with the topic.

Begin with what readers are likely to know and work toward what is new for them.

Make your decisions about balance, proportion, and order when you build your outline. If, however, in the actual writing you deviate from the outline because you discover that "reasons for" is more worthy than "how-to," leave the middle as is but change the beginning accordingly.

Just Try It

Review a document of three or more pages. Determine if: (1) the middle delivers what the beginning promises; (2) the most important elements are given the most space; and (3) the order makes reading easy.

In Your Writing Today ...
☞ **Write middles with balance, proportion, and order.**

24. Writing the Ending, or Leave 'em Laughing

In the theater, the objective is to close a show with the audience doubled over in laughter, drowning in tears, struggling with guilt, burrowed in thought, or bursting into action or song. In other words, don't just leave them sitting there. Close with a punch.

If readers get to the ending of your document, you stand a good chance of winning them over. Don't blow that opportunity with a weak ending, something like, "If you have any questions, call me." The ending is a critical part of the document and in it you should try to achieve specific objectives. Just as the beginning sets the tone, the ending should leave the reader with a particular feeling, thought, urge, or responsibility.

A good ending does one or more of the following:

1. *Summarizes* major conclusions, recommendations, and findings.
   > Regardless of which alternative you select, intensive staff training will be required.

2. States the *significance* of the information already given.
   > What this means, of course, is that all of us will have to reduce expenses and increase sales.

3. Makes a *call for action* (asks for the order).
   > Please give me your comments on this report in writing by Monday, November 12.

4. Conveys a *feeling or attitude* of the writer.
   > I hope this material meets your expectations, for we would like to be a part of this exciting project.

5. Directs attention to the *future.*

   Having absorbed these setbacks of the past twelve months, we have every reason to be optimistic about a successful new year.

Just Try It

1. Write the ending to a memo on office courtesy that makes a call for action.

2. Write the ending to a letter of application that conveys your enthusiasm for working for the company.

3. Write the ending to a report on turnover in your organization that states the significance of your findings.

In Your Writing Today ...
☞ **Don't waste a powerful weapon—the ending.**

WORDING

25. Keep It Simple: Use Conversational Language

Why is it that so many Ordinary Joes become Pompous Pierres when they sit down to write? Conversational language makes reading easier and improves comprehension. Yet some writers insist on using grandiloquent expressions, stock phrases, and bastardized words. Maybe they think their colleagues and clients will be more impressed by "we deem it advisable" rather than "we suggest." But who is impressed by prose they cannot cut through with a machete?

GRANDILOQUENT EXPRESSIONS

GRANDILOQUENT: We were positioned on the proverbial estuary without the proper means of locomotion.
CONVERSATIONAL: We were up the creek without a paddle.

Granted, writing is still more formal than speaking. But its primary objectives are clarity and directness, so why not use the simpler of two alternatives that mean exactly the same thing?

GRANDILOQUENT	CONVERSATIONAL
a multitude of	many
obviate	prevent
differential	difference
counsel out	dismiss
inception	beginning
modification	change

STOCK PHRASES

Writing seems outdated and overly formal when it contains *stock phrases*. Readers tend to lose interest when a letter begins with the worn-out and stuffy "as per your request." And consider these:

STOCK PHRASES	FRESH, CONVERSATIONAL
As per your request ...	As you requested ...
Enclosed please find ...	Here is the ...
Per your memo ...	In your memo ...

BASTARDIZED WORDS

Bastardized words are genuine words made phony by tacking on suffixes: *prioritize, calendarization, orientate, career-wise*. Leave the words in their plain, wholesome form.

Just Try It

Write a simpler, conversational word or phrase for the following:

rectify _____ regarding yours of _____

aforesaid _____ endeavor (verb) _____

strategize _____ ameliorate _____

headcount reductions ____ herewith are _____

effect a change _____

In Your Writing Today ...
☞ **Improve readability with conversational language.**

✔ Stop Wasting Time

One reason many people have an aversion to writing is that it seems to take so much time. Writing a report seems to take twice as long as digging up the information or presenting the ideas orally. That loss of time can be frustrating.

While time is a necessary ingredient in writing, most of us do waste time. The main problem is in getting started. Writing that first paragraph, that lead, that introduction, to set the report off in the right direction, can take an hour or more. Even after you have gathered the information, written an outline, and feel you know exactly what you want to say, the words just don't come out right.

To stop wasting time on a beginning that won't take shape, go to another section of the report, a section that contains facts, figures, steps in a process, or other kinds of hard data that will be easier to get into. You can go back to the beginning later. At that point it will be much easier because the material is more your own after having worked with it. In fact, after writing portions of the report or even the entire report, your views on what to say in the opening may change. So rather than waste time at the beginning, work on another section of the report.

Another opportunity for wasting time comes when you return to a draft you have left an hour or a day ago. Picking up where you left off is like starting over again. To make that return faster, reread what you have written, or at least reread the last few paragraphs or pages. Editing that material will help you recapture your previous train of thought and lead you back into the writing. Ernest Hemingway never stopped his day's writing at the conclusion of a scene or thought. He'd stop in the middle of something, even the middle of a sentence, knowing exactly what words would come next. When he'd come back to the writing, he'd quickly pick up where he had left off. That technique can work for you too.

If you are stuck on a section, leave it. Take a walk; do something different. Leave the project for a day or more and let your subconscious work on it. Sometimes knotty problems seem somehow to unravel overnight or when put on a back burner for awhile. Don't force it.

If a single sentence, paragraph, or concept is blocking your progress, talk it out. Explain to a colleague what you are trying to say. Or picture yourself explaining it to your reader. Listen to yourself or tape your "conversation," and then transcribe it. Chances are you will not only find the words you were looking for, but you will also uncover the flaw in your thinking that was the real cause of the logjam.

In Your Writing Today ...
☞ **Stop wasting time.**

26. Add Vitality with the Active Voice

Powerful business executives earn their reputations as movers and shakers because they make things happen. These are people of action. Yet so much business prose lacks action.

The main reason for actionless writing is the heavy use of the passive voice. In a sentence written in the *passive voice,* the subject is *not the doer of the action.* For example: *The investigation was ordered by the board chairman.* Though the chairman was responsible for the action, *chairman* is not the subject. When a sentence is written in the *active voice,* the subject *is* the *doer of the action. The board chairman ordered the investigation.*

Sentences in the active voice show more action, authority, and motion, and they are more lively and more concise. In addition, in spoken English, the active voice is more natural than the passive voice. Note the differences in the following pairs of sentences:

> PASSIVE: Sweeping *changes were made* by the new president.
> ACTIVE: The new *president made* sweeping changes.
> PASSIVE: The study of payroll deficits *has been completed* by our auditors.
> ACTIVE: Our auditors *have completed* the study of payroll deficits.

The subject does not have to be a person for the verb to be active voice. *Inanimate subjects* can be doers of action.

Prices fell sharply. *Experience reduces* the risk of error. These *requirements demand* increased reserve capital.

The active voice is particularly useful when *defining policy or job tasks.* It forces the writer to identify who is to do what.

PASSIVE: Receipts *are to be collected* at five P.M. (by whom?)
ACTIVE: The *store manager is to collect* receipts at five P.M.

When the doer of the action is *unspecified, irrelevant, or less important than the receiver,* the passive voice is acceptable, even preferred.

DOER UNSPECIFIED: The new requirements are expected to have a great impact on large banks and bank holding companies.

DOER LESS IMPORTANT: The director was served with a subpoena.

Just Try It

Convert these sentences into the active voice *where preferable.*

1. The case will be tried by one of our senior partners.

2. More than a thousand customers were issued inaccurate monthly statements.

3. A decision has not yet been reached by management.

4. Inactive files should be purged from the system daily.

In Your Writing Today ...
☞ **Add action with the active voice.**

WORDING

27. Action Verbs Add Snap

Actions speak louder than words, and action words speak louder than nonaction words. If you do only one thing to improve the vitality of your writing, it should be to use action verbs wherever possible. After writing a draft, see if you can insert a strong, vivid verb for any of the three types of weak verbs: forms of *be*, abstract verbs, and noun-verbs.

TO BE OR NOT TO BE

The verb *to be* and its forms simply mean something exists. *Is, are, was, were,* and *has been* do not show movement or action. Where possible replace these verbs with action verbs. For example:

> TO BE: Working conditions *are* in serious need of a complete review.
>
> ACTION VERB: Working conditions *cry out* for a complete review.

Sometimes you will have to do more than just replace one verb with another; you may have to restructure the sentence:

> TO BE: This procedure *was* not acceptable to the regional office.
>
> ACTION VERB: The regional office *rejected* this procedure.

REPLACE ABSTRACT VERBS WITH CONCRETE VERBS

An abstract verb conveys the *idea* of action. A concrete verb conveys both the idea and its *image*. Use concrete verbs to enliven your writing.

> ABSTRACT VERB: The editing changes *explained* my point more clearly.
>
> CONCRETE VERBS: The editing changes *crystalized* my point.

ELIMINATE NOUN-VERBS

Business writers have a tendency to convert action verbs into nouns, (*formulation, maintenance, development*), forcing the use of a weak verb, often an abstract verb, or the verb *to be:*

NOUN-VERB AND WEAK VERB: The *definition* of user requirements *is completed* in Step One.

ACTION VERB: In Step One we *define* user requirements.

NOUN-VERB AND TO BE: An *understanding* of company policies *is* essential for all new hires.

ACTION VERB: All new hires *must grasp* company policies.

Just Try It

Enliven these sentences by replacing weak verbs with action verbs.

1. We provide assistance to companies without their own PR departments.

2. The opponents discussed the issues for hours without any progress.

3. Computer malfunction was the cause of the erroneous billings.

In Your Writing Today ...
☞ **Use vivid action verbs.**

28. Make References Clear

Anytime readers have to stop to figure out your meaning, or reread a sentence, or chuckle at what the words say though your intent is something else, you jeopardize the success of your document. Unclear references will make the reader do a double-take. Misplaced pronouns and modifiers are the most common causes of unclear references.

MISPLACED PRONOUNS

Be sure readers know who or what the *he, she, it,* or *they* refers to:

UNCLEAR REFERENCE: Jackson sent Oliver the unedited draft because *he* was leaving town that afternoon. (Who is *he*, Jackson or Oliver?)

CLEAR REFERENCE: Because Jackson was leaving town that afternoon, Oliver sent him the unedited draft.

UNCLEAR REFERENCE: The clerks handling the assignment were inexperienced. In the move to the new quarters, files concerning the project were lost and *they* are at a loss as to what happened to them. (Does *they* refer to files? Move *they* closer to *clerks*.)

CLEAR REFERENCE: The clerks handling the assignment are inexperienced. *They* are at a loss as to what happened to the files that were lost in the move to the new quarters.

MISPLACED MODIFIERS

Modifiers are single words, phrases, or clauses that help explain other words. *Keep modifiers close to the words they modify.*

UNCLEAR REFERENCE: The proposal called for a new warehouse *which we finished just on time.* (What was finished on time?)

CLEAR REFERENCE: The proposal, *which we finished just on time,* called for a new warehouse.

UNCLEAR REFERENCE: We are sending our most experienced troubleshooter to examine the site named Patti Lynn.

Readers probably realize that Patti Lynn is the name of the troubleshooter and not the site, but the momentary confusion can be distracting and reduce the effectiveness of the piece and the reputation of the writer.

Just Try It

Rewrite these sentences to make the references clear.

1. When Harry presented the plan to Don he was happy.

2. Edit the introduction to the report and add some graphics to it.

In Your Writing Today ...
☞ **Make references clear.**

29. WHO, WHICH, and THAT

No, this lesson is not a parody of Abbott and Costello's "Who's on first" routine. But the way some people use *who, which,* and *that* is often just as comical. To use these relative pronouns correctly, you have to check two criteria.

Person or Thing

Use *who* only when referring to a person. Use *which* only when referring to a thing or concept.

> Donovan is the *woman who* turned the company around. (Person)
> She gave *direction, which* is what the company needed. (Concept)

You may use *that* with persons, things or concepts.

> The *individual that* made the difference was Fugelsoe. (Person)
> She provided the *vision that* the company lacked. (Concept)

Essential or Nonessential

When your choice is between *which* and *that,* see if the clause introduced by the word is essential to the meaning of the sentence. If the sentence's meaning would change or be unclear without the clause, the clause is essential. When the clause is essential, use *that.*

> The change *that made the difference* began at her first staff meeting.

The meaning of the sentence would be completely different without the underlined clause; therefore *that* is used. Note that when the clause is essential, it is *not* set off by commas.

Which introduces nonessential clauses, those that do not alter the meaning but add information. Nonessential clauses *are* set off by commas.

The company mission, *which Fugelsoe designed*, became law.

PRONOUN	REFERRENT	TYPE OF CLAUSE
who	person	essential or nonessential
which	concept, thing	nonessential
that	person, thing, concept	essential

Just Try It

Correct the misuse of pronouns and commas in these sentences.

1. The investor which bought the company had once been its mail clerk.
2. The Wall Street analysts, that had predicted the takeover, were ecstatic.
3. Total costs which can only be estimated approached $1 billion.
4. An effect which no one foresaw was the wholesale firing of management.

In Your Writing Today ...
☞ Use *who*, *which*, and *that* correctly.

MECHANICS

30. Writing Numbers: Words or Figures?

In business we not only have to *make* the numbers, we also have to *write* them. Business writers can follow the same guidelines for numbers that conventional writers follow.

It is conventional in writing to spell out numbers less than one hundred; use figures for anything over one hundred.

Plans call for hiring *six* more engineers next year.

That will bring the staff to *153*.

BUT

If your document contains many numbers, or you refer to numbers frequently, good practice calls for using figures. The reason for this rule is clarity and ease of reading. It is easier to read many numbers in figures than in words.

Use figures for dates and addresses: April 6, 1989; 310 Main Street

Ordinarily, use figures for:

decimals	3.9 68.5
percentages	22 percent 61%
mixed numbers and fractions	12½
statistics	a decline from 43 to 31
identification numbers	project number 26
volume, chapter, page	Volume II, Chapter 12, page 378
numbers followed by symbols and abbreviations	16 sq. ft. 72 F

72

| exact amounts of money | $9.43 |
| times | 9:15 A.M. |

Spell out numbers at the beginning of a sentence.

> Seventy-two companies responded to the survey.

If necessary, recast the sentence to avoid figures at the beginning:

> AWKWARD: 1988 is the year on which all others are measured.
> RECAST: The company measures each year against 1988.

Just Try It

Assuming the following sentences come from different documents and each contains only a few numbers, change those numbers where necessary to conform to the guidelines given above.

1. Last week 6-month T-bills sold at an average rate of 7 percent.

2. 120 days after the company had been bought, it was sold again.

3. The average family no longer has 2.5 children; the average child now has 2.5 parents.

In Your Writing Today ...
☞ **Follow the rules and common sense in writing numbers.**

✔ Writing's Better the Second Time Around

Time-conscious people like to get the job done right the first time. "Never touch a piece of mail twice," advise the time management experts. So when people hear that good writing requires *revision*, they bristle with the thought of lackluster first efforts and wasted time.

But it is a misconception to think of rewriting as fixing what was done incorrectly the first time. The writer is actually doing different things in the first and second drafts. So think of revision as Act II of a two-act play. The act of writing is such a dynamic process that it requires a second look, which is what *re-vision* means. So what do you look at?

WHAT THE DOCUMENT SAYS

Note that the heading is not "What *I* have said." Try to approach your letter, memo, or report as if someone else had written it. Divorcing yourself from ownership will help you examine the document more critically. One way of achieving this temporary separation is to put the document aside for awhile. Anyone who has ever painted a room knows it is easier to spot "holidays" after the paint dries. So too with writing. In the first draft the writer is so close to the material she "reads into" statements, assuming a statement implies more than it actually does. With a second look, however, she realizes she has to spell out some things.

Or the writer might see the opposite—that an idea was repeated to the point of redundancy. In addition, time allows your subconscious to work on the document, often gathering new information, slants, and arguments. When you make such changes in the *substance* of the document, you will be glad you had the chance—and wisdom—to take a second look.

How It Is Organized

After reviewing the substance, take a look at the skeleton of the draft, its *structure*. Look first at the major divisions and ask if they are in the order that is best for this audience. For example, is *Background, Problems,* and *Proposed Changes* the most effective order? Would the document create a greater sense of urgency and grab the reader's attention if *Problems* came first? Satisfied with the overall structure, then look within each section and ask the same kinds of questions about subsections and the paragraphs within subsections. Dictating the order of material is the ease with which the reader will follow the flow of your thinking and come to your conclusion.

How It Is Said

Finally, you focus on *style*, each sentence's ability to communicate its intended thought. Check to see if each word, especially each verb, best expresses the intended idea. Consider word order. For example, should the word *productivity* go at the end of the sentence for greater impact? You cut most long sentences in half yet strive for variety to create a good reading rhythm. You apply all you know about sentences to each one, so each does its job and fits in with the whole.

Though revision might seem tedious, it can be creative and invigorating, like removing the tarnish from an antique silver tray and polishing it to uncover its intricate design and beauty. You may be amazed at the richness buried in your first draft.

In Your Writing Today ...
☞ **Revise.**

IDEAS

31. Persuasion Starts with the Reader

Almost every piece of business correspondence is intended to persuade. If its overt purpose is not to persuade a customer to do something or think something, its covert purpose is to show the boss how bright its author is. While there are many tricks to persuasion, its most critical element is not a trick at all but the most basic tenet of all writing: Start with the reader. Good writers have mastered the art of expressing what they want to say in ways their readers would like to hear.

John C. Quinn, editor of *USA Today*, illustrated this point of writing to his audience when, with tongue in cheek, he wrote headlines announcing the end of the world as they would probably be written by different newspapers. See if you can match the headline with the newspaper, knowing that each paper shaped the same story to cater to its readers. The papers are *The New York Times, Lawrence (Kansas) Daily Journal World, USA Today, The Wall Street Journal, The Washington Post.**

> STOCK EXCHANGE HALTS TRADING AS WORLD ENDS
>
> END OF WORLD HITS THIRD WORLD HARDEST
>
> WORLD ENDS; MAY AFFECT ELECTIONS, SOURCES SAY
>
> WORLD ENDS; LARRY BROWN STAYS AT KANSAS UNIVERSITY
>
> WE'RE GONE ... STATE BY STATE DEMISE ON 6A ... FINAL, FINAL FINAL SCORES ON 8C

**The Executive Speaker*, P.O. Box 292437, Dayton, OH 45429.

In writing, be conscious of the characteristics of your reader. For example, if he is a conventional, traditional person, emphasize your reliability and tell about proven methods and results. If he thinks of himself as a trendsetter, stress your innovativeness, the unique and leading-edge aspects of your company or suggestion.

Another way to put your writing on the same wavelength as your audience is to recall things they have said. Focus on issues they stressed, their so-called "hot buttons." In doing so you're not just giving back what you heard but rather showing you paid attention to what you heard and have an intelligent response.

Just Try It

1. How would you try to shape your message if your reader were analytical?

2. If she were *authoritarian?*

3. If the management team repeatedly mentioned words like *efficiency, productivity, streamline?*

In Your Writing Today ...
☞ **Persaude by writing *to* your audience.**

SENTENCE STRUCTURE

32. Positioning Ideas in Sentences

Position is important in business and in other areas of life. Position of ideas in each sentence is also important. Think for a moment about how much information can be contained in even a simple sentence; WHO—WHAT—WHERE—WHEN—WHY—HOW (WWWWWH). These parts can often be interchanged in a variety of combinations. For example:

1. The chairman announced a two-for-one split in a letter yesterday to stockholders, signaling the company's financial health.

2. In a letter yesterday to stockholders signaling the company's financial health, the chairman announced a two-for-one split.

3. Yesterday, in a letter to stockholders that signaled the company's financial health, the chairman announced a two-for-one split.

What *is* the best variation? It is the one that is most appropriate for the sentence and its purpose, audience, and context.

The two most prominent positions in a sentence are the first and the last. The last position is the most important because that is where the sentence stops, making the reader pause and reflect on the last word or idea. The first position is prominent because it captures the reader's attention. Therefore, in deciding where to place the WWWWWH of your sentence, consider which component is most important and try to put that *last*. Decide what's next in importance and put that first.

In the sentences above about the stock split, when the writer considers the split itself to be the most important, the idea comes last, as in sentences 2 and 3. If the idea of the company's financial health were most important, that would go last, as in sentence 1. If the timing were relatively important, "yesterday" would be in the first position, as in sentence 3.

Sometimes sentences do not offer this kind of flexibility, but if you think about positioning on every sentence, you will rearrange many of them and give them added vigor. Note that I said "rearrange." Positioning components is done in the revision rather than in the first draft. (Or should I have said "not in the first draft but in the revision"?)

Just Try It

Rewrite these sentences as directed.

1. As last year, increased productivity is the goal for the department this year. (Consider *productivity* the most important idea; *goal* is the second most important idea.)

2. After auditing the department, we find that none of the recommendations we made last year has been implemented and that for the second straight year the department is grossly inefficient. (Most important: *recommendations not implemented;* second most important: *for the second straight year.*)

In Your Writing Today ...
☞ **Position ideas in sentences strategically.**

33. Imagery Makes Readers See the Point

Since the birth of television, America has become a visually oriented society. Even the "Walk/Don't Walk" and "Men's/Women's" signs have been replaced by images. Long before TV, however, good writers were using imagery to enrich and convey ideas. How forgettable Shakespeare would have been had he written things like "experiencing bad luck" instead of "suffering the slings and arrows of outrageous fortune."

Though I'm not suggesting you wax poetic in your business correspondence, you might consider using verbal images to express some abstract concepts. We do this all the time in speech when we refer to results as "the bottom line," to a strategy as "the game plan," to investment capital as "seed money." These expressions are images—graphic representations of ideas that enliven and improve communication.

The easiest way to bring imagery to your writing is through concrete details. For example, instead of "our location provides..." write "our location at the corner of Main Street and Broadway provides..." Instead of "foreign companies are investing in American businesses," write "foreign companies are pouring yen and deutsch marks into American businesses." Think about what is happening or elements that are present, and include the concrete details of the action or setting to create an image conveying the idea.

Another way to develop imagery is to *think of concrete words that have the same meaning as an abstract word*. For example, if you're writing about potential impediments to a project, think of concrete words that mean *impediment*. You will probably come up with words like *barrier, stumbling block, hurdle*. In your document, instead of writing "impediments to overcome," write "barriers to overcome."

Visualize the actions behind ideas. When sales increase, visualize the chart with an ascending line and write about sales "rising." When they decrease drastically, write about their "plummeting." From such basic images you can leap to more figurative images such as "mainstream causes," "escalating costs," and "safety-net features."

Some of these images are so common they may no longer carry a visual content for you. If so, reach for something fresh. You may have to be adventurous, and your readers (or your boss) may be uncomfortable at first. But it was adventurous writers who introduced images like "window of opportunity," "the New York–Philadelphia corridor," and "tunnel vision." They did it, so why not you?

Just Try It

Rewrite these sentences, using a visual image to express an idea.

1. For weeks the department had been trying unsuccessfully to process an unusually high number of requests.

2. The marketing plan includes a series of different ads to reach the public through various media in a concentrated period of time.

In Your Writing Today ...
☛ **Express ideas through verbal images.**

IDEAS

34. You Said It! Using Direct Quotations

Direct quotes are common in sports articles, to give personal viewpoints of the players, or in magazine features in which movie stars expound on their philosophies of cooking or loving or living. But do quotes belong in a business document? Yes. There are at least four good reasons for inserting short quotes in a report, proposal, letter, or memo.

ADD A NOTE OF AUTHORITY

The right quote from an expert in the field can support a position you are espousing. It can tell your reader that your argument is not yours alone, that you've done your homework, and that you know what you're talking about. Supporting this argument are the words of Samuel Johnson: "Knowledge is of two kinds; we know a subject ourselves, or we know where we can find information on it."

BREAK UP BLOCKS OF PRINT

Quotes can break up long blocks of print, giving the reader a breather. Because quotes are set off by quotation marks or often put in a short paragraph of their own, they can offer a respite from page after page of text. Quotes change the pace and pick up weary readers. "The mind profits little from heavy reading."—(Lafcadio Hearn)

DISTINGUISH YOURSELF FROM THE CROWD

The quotes you use can give an insight into your personality as well as that of the person you quote. The quote tells something about the books and papers you read, about the way you assimilate various kinds of information into your work. "Next to the originator of a good sentence is the first quoter of it," said Ralph Waldo Emerson. The quote may distinguish you from the others, who all write in the same conventional way.

SUMMARIZE KEY POINTS

A quote can be an effective clincher, the sentence that puts all of the previously detailed information into a memorable, commonsense statement. For example, you might sum up the benefits of preventive maintenance by quoting that fellow on the TV commercial: "You can pay me now, or you can pay me later." Whether the quote is from a business authority, Shakespeare, the Bible, or some other source, it can crystalize your thinking in a memorable way.

Just Try It

1. Reread some of the reports, letters, and memos you have written. Try to insert an appropriate quote from the intended reader, an expert, literature, or any other source.
2. Thumb through a book of quotations and note those that might have been useful in documents you have written. Jot some down for future use.

In Your Writing Today ...
☛ **Enrich your writing with direct quotations.**

SENTENCE STRUCTURE

35. ... And Everything in Its Place: Misplaced Modifiers and Dangling Participles

"I know a man with a wooden leg named Smith."
"Oh, yeah. What's the name of his other leg?"

That old chestnut may not make you laugh, but it should illustrate a common error in business writing—the *misplaced modifier*. A modifier is a word or phrase that describes another word. A modifier is misplaced when it is placed far from the word it describes.

>MISPLACED: Our intention is to track fluctuations in sales of all merchandise *with the new formula*.

The merchandise does not have a new formula. The new formula describes how we will track. To prevent the misreading, put the modifier nearer the word or idea it describes.

>CORRECT: Our intention is to *track with the new formula* fluctuations in sales of all merchandise.

One particular kind of modifier is so often misplaced it has its own classification. It's called the *dangling participle*. Do you remember that one from school? This is a modifier usually placed at the beginning of a sentence. It dangles because the noun that follows it is not the noun it describes. Here's an example:

>DANGLING: *Organized* properly, *participation* from the trainees will enliven the session.

Organized is the dangling participle here. It's said to be dangling because the noun it is closest to—*participation*—is not the noun it describes. To keep *organized* from dangling, put it closer to the word it describes, *session*. To do that, you'd have to restructure the sentence:

> CORRECT: *Organized* properly, the training *session* will be enlivened by participation from the trainees.

Just Try It

Rewrite these sentences to correct misplaced modifiers and dangling participles.

1. I can count the number of times I've sat before my personal computer on one hand.

2. We assist clients, acting as facilitators, in negotiating contracts.

3. Realizing a bold move was called for, the new product was announced three weeks early. (Hint: Who is realizing?)

In Your Writing Today ...
☞ **Put modifiers near the words they describe.**

What's the Point of Punctuation?

What *is* the point of punctuation, that motley collection of dots and blips harassing writers from grade school to the board room? And *harassing* does not overstate the case. So much energy goes into teaching and learning punctuation that many people come away thinking punctuation *is* writing. I'll never forget the pride I felt showing my recently completed doctoral dissertation to a group of friends. How quickly my bubble was burst when one friend, paging through my weighty tome, said, "I could never write anything like that. I can't punctuate to save myself."

Though punctuation is only a supporting player in writing, it does play a critical role. Without punctuation, written words do not convey the rhythm and pace of speech; or provide the nuances of doubt and certitude; or transfer tone or feeling. Punctuation cannot supply these intangibles fully, but it does bring the written word closer in overall expression to the spoken word.

Take, for example, the pause and the stop, the prime dictators of cadence in speech. Like speed bumps, the *comma* gets readers to slow down, and like the red light, the *period* gets them to stop. Writers can speed a sentence up by constructing it without internal punctuation. Or, when they choose, they can use commas—and other marks—to slow things down.

And what about the afterthought—which often is more important than the forethought? That's conveyed with the *dash*. *Question marks* and *exclamation points* not only ask questions and show excitement, they also put a wrinkle in readers' brows and make their heartbeats quicken. For example: "Just how does this situation affect your company?" "Sales for the quarter rose 29 percent!" How economical punctuation is.

Some punctuation marks are totally misunderstood. Some people—even those who write well—are frightened by the informality of the dash. They totally misuse the *colon* and *semicolon,* cousins in name and appearance only. And many people use the *apostrophe* as cooks would add a pinch of salt—just for seasoning.

Learning to use punctuation can be difficult because some marks have multiple uses. Other lessons in this book teach the most useful and difficult of these uses. But the point is that you can get the best of punctuation if you think of it as a set of little directionals for readers.

In Your Writing Today ...
☛ Use punctuation to capture the intangibles of speech.

36. Variety Is the Spice of Sentence Structure

Back in the days when movies were censored, the censoring body was giving a movie producer a hard time over a scene in which a woman sensuously slithered down a flight of stairs. The dispute was resolved with one simple change: the background music was removed. The censors felt the scene held much less sizzle without the intimations of the music.

Writing has similar kinds of undercurrents to influence readers. One is *pace*. To create excitement, a writer can use a sharp, snappy pace. Or, if the purpose is to convey, let's say, lethargy in the mailroom, a writer can intensify the meaning of the words by setting a slow, languid pace. Short sentences with few details quicken the pace. Long sentences, with multiple details describing a single component, slow the pace.

> SLOW PACE: The interim manager, obviously under self-imposed and external pressure, approached his responsibilities with caution and trepidation.
> FAST PACE: The permanent manager charged ahead from day one.

Length, as you see, affects pacing. But aside from the pace, length is a concern from another standpoint—variety. Too many short sentences in succession give a choppy, telegraphic style that can be annoying. Too many long sentences in succession raise fog and put readers to sleep. Strive for a mix between long and short sentences.

Another undercurrent that influences readers is *word order*. The normal word order for English sentences is subject-verb-object:

> S-V-O ORDER: The *company opened* new *offices* at the rate of four per year over the last five years.

While you can improve clarity by putting most of your sentences in this S-V-O order, you can also create a dull, repetitive rhythm with it. So, to keep your writing lively, change the word order occasionally. You can do that by putting other elements before the subject. For example:

>VARIATION: Over the last five years, the company opened offices at the rate of four per year.

Just Try It

Rewrite this paragraph, creating variety in sentence pace, length, and word order.

>Management holds planning meetings each month. All department heads attend these meetings. Each makes a report on his or her department. Management presents company-wide figures. Management then sets company and department goals for the next period.

Quicken the pace of this sentence to match the idea behind it.

>The new procedure will reduce delay in our responding to inquiries.

In Your Writing Today ...
☞ **Add variety through pace, length, and order.**

MECHANICS

37. Distant Relatives: The Colon and the Semicolon

The colon and the semicolon receive more abuse than any other punctuation marks. Why? These two marks look alike and share a name, but they are very far apart when it comes to function. To master these little demons, ignore their common name and think of their different functions.

THE COLON

Use the colon in:

- the time of day—10:30 A.M.
- the greeting of a letter—Dear Ms. Suczewski:
- a sentence to mean "as follows"—The reasons behind the decision are: the high cost of real estate, the insufficient labor supply, and the inaccessibility to the airport.

The colon also introduces a bullet-point list, as in this paragraph.

THE SEMICOLON

The semicolon replaces the period to join two sentences so close in thought you want to make them one. The closeness of the two thoughts is emphasized when they are together in one sentence.

PERIOD: This office does not intend to draw clients away from other offices in the firm. Nor does it intend to raid offices for staff.

SEMICOLON: This office does not intend to draw clients away

90

> from other offices in the firm; nor does it intend to raid offices for staff.

CAUTION: Use the semicolon to join *complete sentences only*, not dependent clauses or phrases. The semicolon replaces a period, not a comma.

Another use for the semicolon is in replacing the comma to prevent confusion in series of items. Normally commas separate items in a series:

> Management is concerned about slumping sales, increasing overhead, and an aging work force.

But when those items in the series are long and contain commas, the semicolon more clearly marks the distinction between items:

> SEMICOLON: Management is concerned about slumping sales in retail units, wholesale operations, and repair services; increasing overhead, including costs for people, buildings, and equipment; and the aging work force, which now is at an all-time high.

Note how the semicolon sets off one set of items from another. If commas were used, readers would not see the distinction so clearly.

Just Try It

Insert or interchange colons and semicolons where appropriate.

1. Our press release was picked up by the following newspapers; *Graceland News, Valley Shopper,* and *The Sentinel.*

2. The event included speakers from companies such as Coca Cola, IBM, and General Motors, journalists for television, radio, and the general and business press, and attendees from companies throughout the world.

In Your Writing Today ...
☞ **Use, and do not abuse, the colon and semicolon.**

38. Unconventional Techniques Perk Up Business Correspondence

Even when your readers are a captive audience, they can be turned off by dull writing. For example, even if you are writing a memo your boss asked for or a proposal a client requested, don't assume they'll be hanging on your every word. In instances such as these, you may not need as many tricks of seduction as you would in an unsolicited marketing letter, but you still have to work at being interesting. Here are three techniques to help you.

RHETORICAL QUESTIONS

A rhetorical question is one the writer asks but does not expect an answer to. Why ask a question without expecting an answer? So you can answer it yourself, as I'm doing here. By asking the question, you can focus readers' attention and guide them into your sphere of thought. Questions make readers more active participants in the reading process.

Some documents use rhetorical questions extensively, often beginning each section with them, sometimes as subheads. Some writers prefer to ask rhetorical questions sparingly, to draw attention to critical ideas.

ALLITERATION

Alliteration is the term for repetitive sounds in a word or in successive words. You hear alliteration in advertising and promotional names, such as *Better Buy Buick* or *Pru Pack*. You hear it in phrases that catch on, probably because of their sound: *T*win *T*owers, *b*uilding *b*oom, *s*enior *c*itizens. Note that it is the *sound*, not the letter, that causes the alliteration. The sound repetition perks up the reader and makes the words easier to remember. Though you don't want your correspondence to sound like a jingle, it can *b*enefit from some *b*ounce. A well-placed alliterative phrase in your text or in a subhead can break the tedium of conventional prose and highlight an im*p*ortant *p*oint in a *p*assage.

ONOMATOPOEIA

Another advertising technique you might use is onomatopoeia. The term refers to words that sound like what they mean, such as *blast, sizzle, thud*. At first glance you might think this technique has no place in your writing, but in fact business writers do use onomatopoetic words like *glitch, sliding, enmeshed, sluggish, crash*. Many business functions involve action, and where you have action you may have sound. Sometimes that action-sound can be conveyed in one onomatopoetic word. That added dimension might put added snap into a conventional document.

Just Try It

1. Write a rhetorical question introducing reasons for a new procedure you have described earlier in the document.

2. Inject alliteration into the sentence below. You may change words.

    The two glass buildings stood next to each other.

3. Inject onomatopoeia into this sentence.

    After two bad quarters, the company came back strong in the third.

**In Your Writing Today ...
☞ Try the unconventional.**

MECHANICS

39. Dashes and Parentheses: The Interruptors

PARENTHETICAL EXPRESSIONS

In a recent business writing seminar, a participant offered this definition of a parenthetical expression: "A favorite saying of a mother or father." Good guess, but no cigar.

A parenthetical expression is a group of words that interrupts the flow of a sentence or paragraph. Because its thought is tangential to the main idea, it must be set off, usually in parentheses. (Hence, the term "parenthetical expression".) For example:

> SENTENCE: The figures show (see Exhibit II) that sales doubled.
>
> PARAGRAPH: In Peluso's six years as Managing Partner, the firm grew in size and scope. (Peluso had been with the firm only four years when he was appointed MP.) Staff more than doubled, and the firm broke its single focus of commercial property and took on the world.

The information in parentheses is not needed for the reader to understand the main idea. Yet the writer feels the additional information will help the reader. To provide the information but indicate it is not germane to the main idea, set it off with parentheses.

THE DASH

The dash signals another kind of interruptor. The information set off by dashes *is* relevant to the main thought. In fact, the information may be set off by commas, but the dashes call attention to the information more dramatically. For example:

> COMMAS: The company's retail units were sold off, one at a time, to ward off depleting resources from its primary business.

DASHES: The company's retail units were sold off—one at a time—to ward off depleting resources from its primary business.

The dash adds emotion, like a splash of red on a white flyer. It highlights the words it sets off.

Remember that both dashes and parentheses come in pairs, at the beginning and end of the expression to be set off. Don't forget the closer. And note, too, that when typeset, the dash is a single line(—). But with a typewriter or word processor, it's double (--). The double line (--) distinguishes the dash from the hyphen (-).

Just Try It

Punctuate these sentences with parentheses or dashes.

1. To prevent last-minute deadline crunches which are happening on a regular basis we have installed a new schedule for the newsletter.

2. The candidates six in all will be in the office at nine o'clock on Saturday for a full round of interviews.

3. Traffic on Route 197 has increased substantially in the last year. We have been tracking it with funds from a federal grant. Residents see that increase as a torture, but businesses see it as a blessing.

In Your Writing Today ...
☞ **Signal your interruptions.**

40. Avoiding Sexist Language

If you're old enough to remember when the National Weather Service gave all hurricanes women's names, you may be susceptible to using sexist language. Until relatively few years ago, the English language was skewed toward male dominance, especially in business circles. If you are still using many of the expressions reflecting that male dominance, you run the risk of offending a good portion of your readers.

You have to be sensitive, for example, to the slight inherent in *generic* words and expressions such as *mankind, manpower, average working man.* Avoid the slight with nonsexist substitutions like *humanity, work force, average wage earner.*

Years of habitual use of sexist language—whether or not intentional—have fostered problems in constructions using *pronouns.* Grammarians and schools have taught that *he* is acceptable when referring to an indefinite singular person, even when that person could be female. But today that use is perceived as sexist.

SEXIST: A good manager knows the strengths of *his* staff.

Here are some alternatives to solving the sexist pronoun problem.

1. Use *he or she* and its forms. For example:

NON-SEXIST: A good manager knows the strengths of *his or her* staff.

He or she works fine unless you use a pronoun construction in three or four consecutive sentences. For example:

AWKWARD: A good manager knows the strengths of *his or her* staff. *He or she* must also know the weaknesses of *his or her* people. *He or she* cannot expect to get blood from stone or settle for mediocrity.

2. Use the plural.

A solution to this problem of repetition—and to the original problem of the sexist pronoun—is to use the plural:

> NON-SEXIST: Good *managers* know the strengths of *their staffs.* *They* must also know the weaknesses of *their* people. *They* cannot ...

Other solutions to the sexist pronoun problem are:

- Repeat the noun or reword the sentence.
- Use the indefinite "one." (one's staff)
- Address the reader directly as "you." (To be a good manager, *you* must know the strengths of *your* staff.)

These second-tier alternatives may help you out when you've overworked the others or when the others just don't work.

Just Try It

Rewrite this paragraph to avoid sexist language.

> Normally a project of this sort requires at least 180 man-hours. We interview everyone from the salesmen on the road to the girls at the switchboard. We ask each person in the department to describe his job and ask if he has any suggestions for improving his performance.

In Your Writing Today ...
☞ **Prevent insult; avoid sexist language.**

✔ Breaking More Rules

At the risk of encouraging you to abuse the privilege of poetic license, I give you a few more rules to break. But remember, do so only if the writing calls for it.

NEVER WRITE A ONE-SENTENCE PARAGRAPH

In other parts of this book I define a paragraph as a group of sentences developing a single thought. But there is a place for a one-sentence paragraph when used wisely. That is, you limit the paragraph to one sentence *for effect,* not because you have nothing else to say. Because it looks different from the rest, the one-sentence paragraph stands out and draws attention. It is particularly effective in summarizing an argument or position.

Used sparingly, the one-sentence paragraph packs punch.

NEVER SPLIT AN INFINITIVE

An infinitive is a verb preceded by the word *to: to speak, to change.* Good practice tells us to keep the two words together and put modifying words after the pair: *to change immediately* rather than *to immediately change.* Sometimes, however, sticking to the rule creates an awkward sentence: "The training is designed *to prepare* technicians who go into the field *better.*" The sentence is made clearer by splitting the infinitive: "The training is designed *to better prepare* technicians who go into the field."

NEVER USE CONTRACTIONS

Contractions (*isn't, it's, you're*) rarely make their way into formal business documents. They're considered informal, though we use them in business conversations and even in the most formal business presentations. Examine the next marketing letter you receive and note how its "friendly" tone is achieved in part by the use of contractions. With business language moving toward the less formal, more writers are using contractions to make their writing inviting and readable. It's still a matter of taste, though.

Never Write a Sentence Fragment

A grammatically correct sentence has a subject and verb that can stand alone: "The company supports community action." A group of words that does not make sense by itself but is written to look like a sentence is a fragment: "Despite the attempts of the stockholders to curtail the practice." A fragment may be used for effect if it follows a sentence immediately. Because it is not grammatical, the fragment calls attention to itself, emphasizing the words: "The company supports community action. *And always has.*" Do not break this rule twice in a thousand sentences.

In Your Writing Today ...
☞ **Break these rules, but only for effect.**

41. Building a Working Vocabulary

Each of us has at least two vocabularies, a reading/listening vocabulary and a speaking/writing vocabulary. Let's call the latter a working vocabulary. The reading vocabulary is generally the larger of the two, since we probably recognize and understand more words than we use in speech or writing. This lesson shows you how to bring words from your reading/listening vocabulary into your working vocabulary.

The process for building your working vocabulary is rather basic. Each time you read a business publication, make note of words and expressions you find striking, interesting, and effective. Jot them down and try to use one or two in sentences you would write about *your* business. The next time you begin to write, look over your list for words that can convey your thoughts. The words below appeared on one page of *The Wall Street Journal*. They're not fancy and they are recognizable. How many of them are in your *working* vocabulary? When was the last time you used them?

- *enmeshed* in a dispute
- reinforced the *perception*
- *massive* job growth
- *colossal* machine
- *stagnant* earnings
- a take-over *ploy*
- an *aversion* to lawyers
- *cranked out* letters
- *pay homage*
- *disgruntled* owners

Elsewhere in this book I've encouraged you to choose the simple word over the elaborate when both words mean the same thing. I see no contradiction between that suggestion and this one to build your working vocabulary. As you see from the list above, there are many sound, red-blooded words that can strengthen your writing without a taint of affectation.

The act of looking for interesting words in your reading will automatically begin to improve your vocabulary. Keeping a notebook of such words will provide a reservoir to draw from as you write. Once you use these words, they will flow out of your reading vocabulary, out of the notebook, and into your working vocabulary.

Just Try It

1. Write a sentence for each of these words:

   propel, devoid, cites, frozen, impetus, emboldened.

2. List 7–10 words or expressions from your reading that you recognize and find effective but do not use. Use each one in a sentence.

In Your Writing Today ...
☞ Build your working vocabulary.

42. Developing Ideas

An undeveloped idea is like a rosebud that never bloomed. Unless its inherent potential is nurtured, it never reaches fruition. Some writers fail to develop good ideas simply because they don't know how. If the supporting data does not "hit" them immediately, they move along to the next idea, either deleting or burying the original.

To develop an idea, state it in a topic sentence. Then consider the following methods of developing the idea.

DEFINITION: Define a term in your topic sentence.

EXPLANATION: Explain in different words the idea in the topic sentence.

ILLUSTRATION: Give specific examples of the idea.

VERIFICATION: Present facts that prove, explain, and support the idea.

DESCRIPTION: Give details of the idea or situations in which it applies.

NARRATION: Trace events leading to or stemming from the idea.

CAUSAL RELATIONSHIP: Show how one element in the idea leads to another.

CLASSIFICATION: Break a concept into components and explain each as part of the whole.

Some ideas can be developed with a single method; others may require two or more methods. For example, consider this topic sentence:

> In ten years, the company has grown from one unit to a chain of 33.

Depending on your purpose, you might use *narration,* tracing the growth over those ten years. You might use *description,* giving details of certain aspects of it. You might use *causal relationship,* showing how success led to expansion, which led to more success and so on. Or you might well employ a *combination of methods* for developing this idea.

Methods for developing ideas often flow naturally from the idea itself. When they don't, run through the list of methods in this lesson until you hit one that can work for that idea.

Just Try It

Which methods could you use to develop each topic-sentence idea?

1. The project can be accomplished in three phases.

2. I see the domino theory at work in this situation.

3. Management has installed new procedures to combat pilferage.

4. We have made a strong commitment to the cities we operate in.

5. One philosophical concept permeates our entire organization.

Develop one of the ideas above using the methods you identified.

In Your Writing Today ...
☞ **Develop your ideas.**

FLAVOR

43. Readability and the "Fog Index"

Some years ago, in his book *The Technique of Clear Writing*, Rudolf Flesch presented a formula to help writers measure on a quantitative basis the readability level of their writing. He called it his Fog Index, and it is a useful barometer by which writers know when revision is in order.

Flesch's Fog Index, as well as most readability formulas, puts into an equation the percentage of *long words* (three syllables or more) and *average sentence length*. The resulting figure indicates the degree of ease or difficulty readers will find with the piece.

To determine readability of a passage of at least 100 words.

1. Divide the total number of words by the number of sentences:
   total words ÷ total sentences = average sentence length   100/5 = 20

2. Divide the number of long words by the total number of words:
   number of long words ÷ total number of words = percent of long words   16/100 = .16 or 16%

3. Add average sentence length and percentage of long words (drop the decimal point from the percentage of long words):
   20 + 16 = 36

4. Multiply the sum by 0.4 to get Fog Index:
   36 × 0.4 = 14.4

What does this number, 14.4, mean? Flesch applied his formula to many publications and found that no popular magazine had a Fog Index above 12. *Time* and *Newsweek* clocked in around 11; *Read-*

104

er's Digest was around 10. I have applied the formula on different occasions to business periodicals such as *The Wall Street Journal, Fortune,* and *Business Week* and found that on the average the Fog Index was at or around 12. Even the most prestigious business communicators keep the Fog low and readability high, and 12 seems to be the target they shoot for.

So, if they do it, maybe you should too. When you're writing an especially important report, or when you seem to be having more trouble than usual, measure the Fog. If it's much above 12, go back to the basics—short sentences, honest words.

A word of caution: Your Fog Index is not a panacea. You can write all three-word sentences and get a low score, but your prose is likely to be plodding and childish. Don't forget sentence variety—length as well as style.

Just Try It

Calculate the Fog Index on a document you have written. If it is much above 12, try simplifying by shortening sentences and using simpler words. Recalculate the Fog. Then take a deserved walk in the sun.

In Your Writing Today ...
☞ **Keep the Fog low.**

SENTENCE STRUCTURE

44. Reaching Complete Agreement

In business, people can agree to disagree, but you cannot do that in business writing. All subjects must agree with their verbs and all pronouns must agree with the nouns they refer to.

SUBJECTS AND VERBS

Every verb must agree in number with its subject: singular subject = singular verb; plural subject = plural verb. For example:

> SINGULAR/SINGULAR: Bad *weather/has* delayed construction six months.
>
> PLURAL/PLURAL: *Mergers/have* rearranged the rankings of industry leaders.

That all seems simple enough. But sometimes word constructions separate subject from verb, muddying that simple equation:

> INCORRECT: The reason for all *delays have* been the bad weather.

Reason, not *delays*, is the subject, so the verb must be singular.

> CORRECT: The *reason* for delays *has* been the bad weather.

What about *compound subjects?* Two singular subjects joined by *and* take a plural verb; when they are joined by *or*, the verb is singular.

> Procurement *and* distribution *are* major productivity targets.
> Procurement *or* distribution *is* the logical place to start.

When the subject is a *collective noun*, the verb agrees with the intent behind the use of the noun. For example, if by the word *committee* you mean a group acting as a single body, use the singular verb:

SINGULAR: The *committee/meets* on the second Tuesday of the month.

But if you perceive the committee as individuals, use the plural verb:

PLURAL: The *committee/are* in disagreement on the choice of a chairman.

Pronouns

Pronouns must agree with the words they refer to. For example:

SINGULAR/SINGULAR: The press *release* has more than accomplished *its* mission.

PLURAL/PLURAL: Venture *capitalists* have expressed *their* interest in the plan.

Though the writer may be thinking of people in the company, *company* is a singular noun, and takes a singular pronoun.

INCORRECT: The *company* asked that we address *their* cash concerns first.

CORRECT: The *company* asked that we address *its* cash concerns first.

Just Try It

Correct errors of agreement in this paragraph.

The meeting of company stockholders were held on Monday. The Chairman and the management team was present to answer questions, most of which revolved around rumors of plant closings. The Chairman or an appropriate designee were evasive in discussing the issue. Management apparently will not tip their hand prematurely.

**In Your Writing Today ...
☞ Be agreeable.**

45. Fragments

Fragmentation is one of the major causes of stress in modern society. When we give pieces of ourselves to so many endeavors, without any continuity among them, we can lose a sense of internal wholeness.

Sentence fragments cause a similar kind of disorientation, though certainly on a different plane. *A sentence fragment may look like a sentence but lacks completeness of thought.* Readers become puzzled or distracted, looking for more or feeling as though they have missed something.

We sometimes create a fragment by tacking an *afterthought* onto a valid sentence. For example:

> FRAGMENT: After due consideration, we decided to move the Information Center. *Because we think the results dictate a change.*

The afterthought has a subject (*we*) and a verb (*think*), but standing alone, this group of words does not make a complete thought.

> CORRECT: After due consideration, we decided to move the Information Center, because we think the results dictate a change.

A less obvious error can occur in a larger fragment, one that has clauses with subjects and verbs, giving all appearances of a sentence. But unless there is a subject/verb combination giving the *main idea*, there is no sentence. For example:

> FRAGMENT: Though our Boston office was eager to house the function, and in fact gave it a reasonably good effort.

This group of words is not a sentence. It does not express a complete thought. Reading that fragment, you wait for the main idea, *though our Boston office was eager ...* To correct the fragment, complete the thought.

CORRECT: Though our Boston office was eager to house the function, and in fact gave it a reasonably good effort, the results suggest we move the function to Boise.

Just Try It

Rewrite these sentences, correcting the sentence fragments.

1. The direct-mail piece will focus on a case study. A project we completed successfully for a client.

2. When you asked us to propose on this subject, back in September, to research the market to determine the feasibility of your introducing your X224 to the industry.

3. Your letter called for a quick, summary-level response. We sent it.

In Your Writing Today ...
☞ **Avoid sentence fragments.**

✔ About Form Letters and Boilerplate

Are you turned off by bulk-mail letters that greet you with "Dear Occupant"? The disinterest you feel is probably what your business associates feel when they receive your form letters or proposals laden with boilerplate. Like the last child in a family of six, these people may very well resent your sending them hand-me-downs.

Granted, form letters do save time and can be useful when the message is routine. Or if a person is really interested in the manual or prospectus you are sending, you are safe in sending a form letter that says "Here is the material you requested." You've made a favorable impression by sending the goods. But if the letter is supposed to move the reader to action, it should not only look customized, it should also be designed for a specific purpose. One letter can go out to many people in a "single mailing," but using the same letter over and over is like repeatedly serving last night's meatloaf.

The real problem with form letters is that they live too long. If your business needs the efficiency form letters can give, review them every six months. See if they still cover the elements that need to be covered and if they are still accurate. Even if the message is still valuable, it's a good idea to revise the document to keep the message fresh.

Boilerplate does have a place in proposals and reports requiring an overview of the company. The problem with boilerplate, though, is that it can get heavy, dull, and rusty over time. Minor but important changes in the company may not have been made. Aspects of the company's history or organization that could be more important to one client than another will not get the proper emphasis. And the use of the boilerplate in one section encourages the tendency to write the whole proposal in cut-and-paste fashion. The client soon recognizes he's been given a "one-size-fits-all" proposal, which does not win engagements or clients.

Even when writing the most routine letter or memo, resist the urge to consider this transaction the same as a hundred others. It may well be, but try to treat each document as something new. Doing so will sharpen your writing. It may also freshen your approach to your work. But most important, an original, tailored document conveys the impression that you care about your work—and your reader.

In Your Writing Today ...
☞ **Be fresh.**

46. Drawing Comparisons

In the days of King Arthur and the Round Table, the White Knight often entered combat against the purveyors of injustice to defend those unable to defend themselves. Most literate people are at least vaguely familiar with the legends of the White Knight, so the term and the idea became a useful metaphor in business parlance to describe the activities of one company defending another from a hostile takeover.

A *metaphor* is a comparison of two seemingly unlike subjects used to clarify one of them. Arthur's White Knight and a modern-day business are two different subjects, yet they have a significant point in common—coming to the aid of the defenseless. That point of commonality helps clarify the function of the business in the takeover battle.

Comparisons in business writing help clarify a role, a position, or a situation by expressing the idea *in another way*, a way other than the literal. These comparisons are most often metaphors and similes. The simile uses *like* in the comparison; the metaphor does not.

SIMILES: Your equipment is *like* a six-fingered glove: it offers much more than I need or could ever use.
Our auditors were treated *like* ants at a picnic.

METAPHORS: His lucrative contract was a set of golden handcuffs, keeping him at the company much longer than he would have liked.
The office ran with the same precision of a crack military unit, but without the same sense of loyalty.

For similes and metaphors to be most effective, they should be short and draw comparisons to familiar things so that the point of comparison in each case can be grasped easily. And be sure you express each comparison in just a few words.

Just Try It

Write a sentence using a metaphor or simile to help explain each of the following. Identify the method of comparison used in each sentence.

1. the department's poor attitude

2. a manager's boring memos

**In Your Writing Today ...
☞ Draw sharp comparisons.**

47. Collecting, Selecting, and Ordering Information

A lengthy report requires a lot of information, often gathered from different sources. A systematic approach to collecting, selecting, and ordering that information produces a better document.

COLLECT AND RECORD RELEVANT INFORMATION

Usually information can be gathered from a variety of *sources*: co-workers, clients, observation, original thoughts, questionnaires, newspapers, dreams, experience, and previous documents of the same type. The information may come in many forms: oral comments, fact sheets, graphs, charts, illustrations, internal whisperings, opinions.

In the early stages of the writing process, you gather relevant material from all the available sources. You discriminate merely on the basis of relation to your subject and *potential* usefulness.

Some people collect this information on assorted tearsheets, photocopies, message slips, napkins, computer printouts. A more orderly place to record data is on the *note card*. Writing one note to a card, you are putting the information in the form that is easiest to work with. When you start sorting out the data by topics, you can easily form small piles of note cards containing related information. This sorting process is more difficult when you have a variety of topics on a single sheet of paper.

SELECT THE MOST PERTINENT INFORMATION

Review the information for deeper understanding and relationships. Assign a *topic heading* to each piece of information. This topic heading would be a classification or category into which the note or fact would fall, such as *history of the company, objectives, timeframe*. Obviously, cataloging data is more difficult if you have four or five different topics on a single page.

In labeling information this way, you may find that some of your information no longer seems pertinent. Ruling out information is just as important as cataloging it, so just put that information aside.

ORGANIZE WHAT YOU'VE SELECTED

Pull together into piles all notes with the same topic heading. Reread the notes within each heading and subdivide these further. For example, *history* might be further broken down into the subheads *origins, major growth periods, years under Johnson.* Working on sheets of paper, you would rewrite all information with the same topic heading on a single sheet or successive sheets. Then you'd assign subheads to each item. Working with note cards, simply sort cards into piles by subheads.

In a real sense, these piles are the outline for your report.

Just Try It

Plan your next long document "by the book." That is, follow the directions given here and use note cards. The process may seem tedious at first, but ultimately you will save time and write a better paper. Trust me.

In Your Writing Today ...
☞ **Research and plan before you write.**

48. Common Errors in Word Usage

Close only counts in horseshoes. In business writing, use the right word, not one that looks like it or means roughly the same thing. Here are some words whose appearance and meanings cause more trouble than they should. So let's master them once and for all.

affect, effect. *Affect* is the verb; *effect* is the noun.
The promotion did not *affect* sales. But it did have an *effect* on the traffic in the stores.

Effect can be used as a verb to mean *to bring about*. "New managers seek to *effect* change." I recommend your using *to bring about*, not *effect*.

affect, impact. *Impact* is a noun and should *not* be used as a verb.
INCORRECT: "The merger *will impact* all departments."
CORRECT: "The merger will *affect* all departments."
or: "All departments will feel the *impact* of the merger."

compose, comprise. *Compose* means constitute, or make up: Three divisions *compose* the entire company. *Comprise* is the reverse idea. "The company is *comprised* of three divisions."

due to, because of. *Because of* is the form preferred by three out of four dentists who chew gum, basically for grammatical reasons. Without getting into those reasons, let's just say use *because of*.

flout, flaunt. *Flout* means to scoff at, as one who "*flouts* the law." *Flaunt* means to display ostentatiously, as "The new partner *flaunted* his recently attained status."

head, head up, meet, meet up. Down with the *up*. People *head* committees. People *meet* other people.

imply, infer. The speaker or writer *implies*; the listener or reader *infers*.

it's, its. The apostrophe here does *not* signal possession.
it's = it is. "*It's* an unrealistic approach."
its = possessive. "*Its* philosophy is weak."

less, fewer. Use *fewer* for numbers, *less* for amounts, degree or value.
Lipex has *fewer* locations but more sales reps.
Jones shows *less* initiative than Smith.

Just Try It

Correct all errors in word usage.

What we imply from your letter is that its our responsibility to head up the Committee for Restitution. The Committee will be composed of less than six people. Its objective will be to impact the attitudes of all citizens we meet up with during the year.

In Your Writing Today ...
☞ **Choose the correct word.**

49. Jargon: Not for "Outside" Audiences

You might be perfectly comfortable with "seed money" and "greenmail," but what do "swipe" and "woodshedding" mean to you? If you were a barbershop-harmony enthusiast, you'd know that "swipe" refers to voices sliding up or down at the end of a line, and that "woodshedding" is four guys improvising harmony to a song.

Just about every business, hobby, or field of interest has its own vocabulary—its own jargon. People within the group use their jargon interchangeably with the vocabulary the rest of us use. The problem arises when they forget how proprietary that vocabulary is and start using it with "outsiders." It's like speaking a foreign language to the uninitiated. No matter how loudly you shout, the words mean nothing.

In writing a letter, report, proposal, or other document, consider your readers. If you know for certain they are card-carrying members of the same business "fraternities" you belong to, feel free to use the jargon. But if you suspect that some of your readers might have trouble with your specialized vocabulary, either stick to standard English or be sure to define any jargon you use. For example, your report may be directed at an MIS (Management Information System) department, so you would assume those readers would know the MIS jargon. But if someone outside that department, such as the Executive VP of Purchasing, will also read the report, be sure that reader can stay with you.

While jargon simplifies communication among "insiders," it confuses and excludes the rest of us. To ensure clarity, consider the effect of your jargon on your readers—all of them.

Just Try It

1. Review ten pages of material you have written on the job. List terms that qualify as jargon, vocabulary not recommended for "outside" audiences.
2. List four readers who *might* have problems with these terms.

In Your Writing Today ...
☞ **Don't jar readers with jargon.**

MECHANICS

50. Abbreviations: The Long and Short of It

I'll try to be brief in talking about abbreviations. A good guideline to follow is to use abbreviations only *rarely* in the text of a document. Spell out words like *street, avenue,* and *company* in a letter (though you might abbreviate them in an address).

Use abbreviations, however, for titles before and after names.

BEFORE: Ms., Dr.   AFTER: Ph.D., J.D.

Most titles that people carry use periods in the abbreviations. The opposite is true, though, for most government agencies: IRS, SEC, FDA.

For some reason, a number of abbreviations for Latin expressions are still hanging on in English. Here are the most common.

i.e. (id est) means *that is*
e.g. (exempli gratia) means *for example*

Not all readers, especially younger ones, know what these abbreviations mean. Rather than run the risk of confusing your readers, why not just use the English—*that is* and *for example*? If you insist on the abbreviation, be sure to place periods after the letters and set the abbreviation off with a comma before and after:

Intangible benefits, e.g., improved morale, also flow from "pay for performance programs."

Etc. (et cetera) is another Latin abbreviation, meaning *and so forth.* Use it sparingly, and only when the "and so forth" items are obvious. Don't expect readers to fill in information they do not have.

WRONG: We should meet to discuss conditions of the sale, etc.

120

CORRECT: This meeting will focus on administrative matters only, such as time sheets, expense reporting, billing procedures, etc.

Sic means *thus it is* and is used primarily when quoting someone and the quote contains an error. *Sic,* placed in parentheses after the error, indicates that the error was in the original. For example:

The stockholder wrote: "Isn't it time the COE (*sic*) started acting like a Chief Executive Officer and not an administrator?"

Just Try It

Make the appropriate changes in the use of abbreviations in this paragraph. Where the abbreviation is inappropriate, remove it.

> We received a letter from Polly Esther, Ph.D., a mgr. working in the Consumer Relations Dept. for the E.P.A. Her letter alludes to our procedures for dumping waste, etc. Ms Esther warns that we are in violation of several federal regulations. Some of these regulations, i.e., those on toxic waste, carry severe penalties.

In Your Writing Today ...
☞ **Use abbreviations sparingly and correctly.**

JUST THINK ABOUT IT—10

✔ The Written Proposal Is a Selling Statement

So much new business rests on written proposals. An effective proposal is a selling statement, an *act of persuasion*. To win, the proposal must offer the client the best solution to the problem and provide evidence that you are the best resource to apply that solution. Too many proposals fail because they ignore the client's problem or need. They run on forever about the merits of the seller and never get to what the client is interested in—the problem in his or her company.

To focus on the client's needs, study the Request for Proposal (RFP) and respond to it. Take notes of your conversations with the client and underscore the "hot buttons." Make clients feel you listen to them and understand their problem.

Not long ago I was asked to edit a long proposal. Its writers said they were having trouble "organizing" what they wanted to do for the client. Yet the skeleton for that organization had been in their possession all along in the form of a letter from the client outlining exactly what he was looking for. That letter, like most RFPs, made a perfect outline for our response.

Sometimes the client has not yet pinpointed the problem, which makes your job more difficult. In that case, work a little harder to unearth the problem. Chances are, the proposal that helps the client identify the problem will win the engagement.

After you've identified or stated the problem, design a strategy to solve it. Study all alternative solutions and select the one that enables you to use your strengths to achieve that end. Help the client see how this solution is tailored to this specific problem.

Get to the point with detailed, specific statements. Unsupported generalizations impress no one; neither do empty affirmations about

Some of the information presented here comes from *The Consultant's Guide to Proposal Writing* by Herman Holtz (New York: John Wiley & Sons).

how wonderful you are. Express your pluses and advantages over the competition *in terms of how these qualifications meet the immediate needs of the client.* Explain an approach and submit a work plan the client can follow logically to the meaningful solution. Let your details and evidence of research convey your expertise.

The writing should be clear and direct. Don't indulge in purple prose. Don't hedge or overqualify statements. Don't fill twenty-five pages if ten will do. Use nouns and verbs, not adjectives and adverbs; and use the active voice, not the passive. Use graphics, subheads, underlining, and such to highlight your approach to the client's problem. And remember, neatness and punctuality count.

In Your Writing Today ...
☞ **Focus proposals on the client's need.**

SENTENCE STRUCTURE

51. Run-Onnnnnnnnnns

The *run-on* is an aptly named error. It refers to a sentence that, from a grammatical standpoint, runs on longer than it should. Some sentences can be called "long-winded," "exasperating," "rambling," and "confusing," but if they are grammatically correct, they are not run-ons.

The *run-on*, essentially, is the forced merger of two sentences without the proper punctuation. For example:

> RUN-ON: Peter Milton announced his intention to buy back a good portion of the company stock <u>he is obviously concerned with takeovers.</u>

The underlined words form a sentence of their own and cannot be tacked on to the preceding sentence without the proper punctuation.

> CORRECT: Peter Milton announced his intention to buy back a good portion of the company stock. He is obviously concerned with takeovers.

Run-ons often develop when the writer's mind is racing ahead and moves on to the next thought before giving complete expression to the previous one. Sometimes the writer uses insufficient punctuation—a comma only—to join two sentences, and that creates a run-on.

> RUN-ON: The salary was not enough incentive to attract people to the added responsibility, no one applied.

Correct the run-on by reinforcing the comma with a conjunction, such as *and* or *but*, or by replacing the comma with a semicolon or period.

> CORRECT: The salary was not enough incentive to attract people to the added responsibility, and no one applied. (*comma and conjunction*)

The salary was not enough incentive to attract people to the added responsibility; no one applied. (*semicolon*)

The salary was not enough incentive to attract people to the added responsibility. No one applied. (*period*)

Just Try It

Correct these run-ons.

1. The job entails providing many documents simultaneously it requires someone with good organizational skills.

2. The individual in this position must realize it is a service function, he or she must be receptive to requests for help.

3. Good service is our business business is good.

**In Your Writing Today ...
☞ Avoid Run-Onnnnnnnnns.**

IDEAS

52. Discriminate and Delete

All writers struggle with what to put into a memo, letter, or report, but too few struggle with what to take out. The reason? They don't take out anything. Many potentially effective documents stagger into incomprehensibility because of the "excess baggage" they carry. Being a discriminating editor of one's own writing requires cold objectivity, and may mean deleting favorite lines or expressions, which is like asking a parent to exile a son or daughter. But less adornment and distraction will enable the essence of the centerpiece to shine through.

To become proficient at deletion, save this task for late in your revision process, after you've gone through two or three drafts "building up" your argument. When you feel you've said just about all you have to say, read the document slowly from start to finish. This complete reading will make duplications more obvious to you. As you read, put a mark (✐) next to each passage you think may be redundant or adds little or nothing to a thought. Then go back and cut.

Look upon each word, sentence, paragraph, and idea as a worker in your employ, and fire every one not carrying its own weight.

One artificial aid you can use in the deletion process is to "kill off the widows." A "widow" is a single word or two at the end of a paragraph that necessitates a new line. To save valuable space in publications, editors find ways to reword a sentence in the paragraph to "pick up" that extra line. When you start killing off widows, you can become quite adept at deletion and compression.

In writing this book, for example, I was limited to forty-two lines per page, and killing widows helped me pick up many lines. What pearls of wisdom I had to deny you, but most of those gems were redundant or irrelevant. In writing, less is often more.

Just Try It

Delete unnecessary words and ideas from this passage.

Working a convention to develop leads and business takes experience and skill. Training in correct behavior can provide some of the skill. For one thing, one must arrange the booth attractively to draw conventioneers to it. A well-laid-out booth will go further and pull those conventioneers passing by onto the carpet and into the booth. The sales rep working the booth then must know how long to let the person browse before approaching, and how to offer assistance. The wrong approach will cut off the visitor's interest and cause him or her to leave immediately.

In Your Writing Today ...
☞ **Remember that less is often more.**

53. More Troublesome Words

Lesson 48 has proven to be a popular lesson, so here are some more troublesome words to master.

regardless, irregardless. There's no choice here at all. *Irregardless* is always incorrect, regardless of how often you hear it.

orient, orientate. Regardless of where people are coming from or going to, we *orient* them. *Orientate* is pretentious.

relevance, relevancy. The correct form is *relevance*.

Mean, average, median. *Mean* and *average* are synonyms, referring to the sum divided by the number of components. "The *average* salary is $49,250." *Median* is in the middle, the number with as many components above it as below it. "The *median* salary is $52,000."

verbal, oral. *Verbal* connotes reducing ideas to words, either written or spoken, as in: "The testimonial gives *verbal* expression to our feelings." *Oral* refers to spoken communication. "An *oral* agreement was sufficient for both parties."

over, more than. *Over* refers to space, as in "*over* the city." *More than* is used with numbers, as in "*more than* 300 people attended."

principle, principal. A rule or truth is a *principle*, as in "The *principle* behind the policy is self-determination." *Principal* is usually an adjective meaning main or primary, as in "the *principal* advantage."

unique. The word means *one of a kind*, so a thing or an event cannot be *very* or *somewhat* unique. It's just *unique*.

up, raise. *Up* is an adverb. Don't use it as a verb, as in "to *up* the percentages." For this idea, use *raise*.

who's, whose. *Who's*, like *it's*, is not a possessive. *Who's* means *who is*: "Randy is the one *who's* about to take over the company." *Whose* is the possessive: "It was Randy *whose* contribution made the difference."

who, whom. *Who* refers to one doing an action; *whom* to the recipient: The employee *who* offers the best suggestion receives the award. Leslie was one person in *whom* I had complete trust.

Just Try It

Correct the errors in this passage.

The meeting was held to orientate staff of the acquired firm. New management explained that as people of principal, the negotiating parties had reached only a verbal agreement and that irregardless of rumors, over 75 percent of the staff would be retained. Any employee who's job was eliminated would receive a very unique form of severance.

In Your Writing Today ...
☛ **Make less trouble for your readers: Use the right word.**

54. Proofreading Pays Off

Proofreading is one of those mundane chores that annoys the truly imaginative writer like you and me. Yet it is one we can't ignore because a misspelling or an omitted word can distract readers or cause them to think of the writer as careless and slipshod, one they'd rather not trust with important details, such as their business.

The best way to proofread is to *look for different things in each reading*. For example, on one pass look only for *misspelled words*. If you find yourself being drawn into the crystalline thought of your prose, scan each page from the bottom up so you can focus on spelling. Too often we overlook the obvious, inadvertently misspelling the name of the person or company we are sending the letter to. And don't rely entirely on computer programs to catch all spelling errors. These programs will not "flag" the wrong word if it's spelled correctly, and will allow things like *pubic sector* when you mean *public sector*.

On another pass, look for *punctuation errors*—an end mark at the end of each sentence; the second comma, dash, or parenthesis; colons and apostrophes in the right place. Look, too, at abbreviations and department names for upper- or lower-case and for commas and periods.

On one pass, check for *consistency*. If you are using all caps in section titles or underlining subheads, be sure all of them are treated the same. Make sure section titles and subheads carry the same wording as in the table of contents. And while you're looking at the TOC, check the page numbers of titles and subheads against the actual page numbers in the text.

When *checking for changes* you have made or authorized, look not only at those but also at the words and sentences around them. Sometimes, particularly with word processors, too many or too few of the words around the change are deleted, creating new errors.

Like the finishing touches on a gourmet meal, the adjustments you make through proofreading can strongly influence the overall impression your writing has on your readers.

Just Try It

Proofread this passage and correct all errors.

Perhaps we're not had many sales with this service line because there is not need for it. Are we trying to sell something people don't need or want. Are we pushing the technology without showing what business problems the technology solves? I feel—as I have all along that thisservice is of value to clients and will be profitable for us

In Your Writing Today ...
☞ **Proofread with a purpose.**

55. Writing Directions

A common task for many business writers is the "directions memo." You may at times have to instruct people on how to implement a new procedure or use a new marketing tool. You may have to give guidelines for using a new computer program. Or perhaps it's your job to tell people how to file for medical insurance or to contribute to United Way.

In giving directions, be concerned about how you refer to your readers. Some writers always use the informal *third person:*

> To prepare an OIS plan, the office manager should determine needs.

Using the third person is acceptable, and perhaps even preferable, when others besides the individuals named (in this case the office manager) will receive the memo. But if the memo goes to office managers for their use only, why not be more direct? Why not write to them as you would speak to them, in the *second person?*

> To prepare an OIS plan, *you* should determine needs.

Readers feel an intimacy with the writer when they are addressed directly as *you* and are drawn closer to the subject.

If you are reluctant to get that personal, you can speak directly to the reader, leaving out the pronoun *you,* and merely implying it:

> To prepare an OIS plan, determine needs.

Be consistent and stick to the form you have chosen:

> INCONSISTENT: The *Office Manager* should investigate all alternatives for meeting those needs. *You* should consider equipment compatibility.

Another key point in writing directions is to use *active voice verbs.* Directions tell people what to do, so phrase them with active verbs.

PASSIVE VOICE: An RFP *is prepared* and sent to vendors.

ACTIVE VOICE: The manager *prepares* an RFP and sends it to vendors.

The active voice not only complements the message, but it also forces you to *specify who is to perform the action.*

Put your directions in a *logical sequence,* and keep each step as brief as possible. But where it would help, give a few words on *why* the step should be done this way or that. Understanding *why* can often help people with the *what.*

Just Try It

Rewrite the directions that follow, using the suggestions in the lesson.

To have the mailing list updated, these steps should be followed:

1. The current mailing list should be reviewed for errors.

2. Names no longer relevant to our purposes should be deleted.

3. The edited lists should be returned to me.

4. Modifications on names, addresses, titles, etc., should be made.

5. Names not on the list that are wanted on should be added.

In Your Writing Today ...
☞ **Write directions readers can follow.**

✔ Writing on a Word Processor

The computerized word processor is a useful tool for writing. I mean *real* writing, not just transcribing, as was done in the "Typing Pool" and is now done in "Report Processing." The word processor can help us greatly in transferring ideas on paper from one mind to another.

The word processor helps improve writing because it gives writers extreme flexibility for *revision,* a key step in the writing process. It enables the writer to add and delete with ease, and to rearrange words, lines, and entire passages. The writer can then "test" the new arrangement and if it doesn't work just slide things back where they were. Writers of course, could always do these things on paper, but with a lot more fussing, marginal insertions, and painstaking cutting and pasting. Because the word processor can help us revise, it makes a major contribution to real writing. Plenty of software packages make revision easier by analyzing a document for "high fog," pointing out long sentences, wordiness, vagueness, and so on. We still have to decide what to do with these problems, but the computer's objectivity and capacity for detail can set us in the right direction.

Unfortunately (you knew this was coming), many people who write on a word processor waste its revision capabilities. Once they pound out the first draft, that's it. We know, however, that even though the product may look good, too often it's far from "finished." But that isn't the word processor's fault.

Many writers do use the *spelling-check* features in word processing programs to simplify proofreading. This, too, is a great aid, but don't rely totally on computerized spell checkers because they point out only misspelled words. If you have used the wrong word but spelled it correctly, the computer will not highlight it.

Another problem with the word processor is that it causes writers to ignore what they know about *pre-writing*—about thinking out, planning, and outlining a document. Many writers go right to the word processor and start typing out sentences, without going through the necessary preliminaries that they once did on paper. They "input" paragraph after paragraph and print it out. End of job. The ease of getting from start to finish gives writers a false sense of assurance. There are programs to help writers plan, but few people use them, maybe because they're cumbersome.

Of course, I really can't blame the word processor for our not planning or revising any more than I can blame a copy of *War and Peace* for being used as a table support. The word processor is a tool—a powerful tool—but only if the writer uses it well.

In Your Writing Today ...
☞ **Use the word processor wisely.**

56. More on Sexist Language

Tradition has taught many of us—men and women—to use certain male-oriented words that today send out chauvinistic signals. If you habitually use words like *foreman* and *stewardess*, break that habit and use more generic terms like *supervisor* and *flight attendant*. Old habits may be hard to break, but you (regardless of your sex) don't want to offend colleagues or clients who resent language that implies male superiority. To break those habits, keep these language traps in mind.

1. One subtle offense comes from mentioning the sex of the person *only* if the person in question is a woman. For example:

   SEXIST: The consultant, *a woman engineer*, recommended major changes.

If you would not write *a man engineer* in that same sentence, then do not write a *woman engineer*, as if it were such an astounding exception.

2. Do not use physical characteristics in describing women in business when you would not use them in describing male counterparts.

   SEXIST: The litigation team was headed by Valerie Reed, an *attractive* and brilliant attorney.

Unless you'd write *handsome* if the attorney were male, leave out the *attractive*. It plays on an outdated notion of women.

3. Do not use *lady* or *girl* when you legitimately do need to refer to sex of individuals. Instead, use *woman*. For example:

   SEXIST: *Girls* outnumber men in our office by six to five.

Girl refers to a young woman and would be the equivalent of *boy*. *Lady* connotes white gloves and dainty tea cups and should be used only in situations calling for these.

4. Do not denigrate women who work at home by using "work-

ing women" to refer only to women who have salaried jobs. Instead, use expressions such as *women who work outside the home* or *salaried women*.

5. Do not use job titles that specify sex when the job is often done by people of either sex. Use *mail carrier*, not *mailman*, and *sales representative*, not *salesman*.

Just Try It

Rewrite this paragraph, removing the sexist language.

> Normally a project of this sort requires at least 180 man-hours. Our project director, a woman management specialist, interviews everyone from the salesman on the road to the girls at the switchboard. To comply with EEO regulations, we ask the ladies if they feel they are treated fairly. We ask the group to identify a spokesman who will report on department-wide complaints.

In Your Writing Today ...
☞ **Don't insult; avoid sexist language.**

57. Setting the Right Tone

"I don't like your tone." This mother's warning has importance for business writers because it reminds us that it's not only what you say, but *how* you say it. Think about the different tones you can use in saying, "Are you going?" That question can be posed in a tone that is threatening, whimsical, inquisitive, or irritable.

Writers should be conscious of tone for two reasons. First, if you inadvertently convey the wrong tone, you can confuse and even alienate the reader. For example, if you are in a whimsical mood and begin a memo with some harmless joke, your readers may think you are taking a serious issue too lightly. A second reason for setting the right tone is to use it to your advantage in presenting readers with your feelings as well as your ideas.

How do you inject tone into writing? First of all, with *word choice*. Conversational language conveys a much lighter tone than formal language. Note the difference in tone in these two openings:

FORMAL: Enclosed is the document requested by your office.

CONVERSATIONAL: Here's the document you asked for.

Sentence structure can also affect tone. Longer, complicated sentences create a formal, white-shirt-only tone. Shorter, simpler sentences usually do just the opposite. But sometimes a series of short sentences can carry anger: "This situation must stop. Now. Or else."

Third, *figures of speech* will influence tone. Metaphors and similes, not too common in business writing, can convey a fresh enthusiastic tone. When you write of "opening a new window of opportunity," you convey more enthusiasm than when writing "there is significant opportunity."

Fourth, *rhythm* influences tone. To determine the rhythm in your writing, read it aloud and listen for the natural ebb and flow, for when the pace quickens and slows, for the words that are stressed because of where they're placed.

Just Try It

Rewrite these statements to achieve the tone indicated.

1. As per your letter of 7/2, the following decisions have been made. (Rewrite with more informal tone.)

2. Before the sheriff comes to throw us out and bar the door, let's see what we can salvage. (Rewrite in less whimsical tone.)

Write a short memo announcing to your department that the company is being sold. Strike a serious yet nonthreatening tone. Indicate that people may lose jobs, but try to assuage people's fears.

In Your Writing Today ...
☞ **Set the tone appropriate to your audience.**

58. Advice on Adverbs and Adjectives

The best advice there is on adverbs and adjectives is try to avoid using them.

To refresh your memory, *adverbs*, most of which end in *-ly*, modify verbs, telling *how* or *when* an action takes place: "argued *incessantly*," "called *periodically*." Some adverbs do not end in *-ly* and often tell *where* an action takes place: "fuming *inside*," "sitting *there*."

Adjectives modify nouns, qualifying the noun in some way: "*prudent* decision," "*huge* deficit," "*administrative* matters." As you can see, adjectives can convey many features of nouns, such as *qualities* (prudent), *size* (huge), or *type* (administrative).

These seem like upstanding English words, so why do I advise that you avoid them? Because, like bad companions, they are the occasion of sin. Consorting with them can cause you to get sloppy in your choice of nouns and verbs—the backbone of a sentence. Shoring up with adverbs and adjectives is never as strong as rooting in with nouns and verbs.

> **WEAK VERB AND ADVERB:** Sales fell *sharply* last quarter.
>
> **STRONGER VERB:** Sales *plummeted* last quarter.

The same principle applies with adjectives. The right noun packs twice the potency of an approximate noun and an adjective.

> **WEAK NOUN AND ADJECTIVE:** Our task was to remove *large piles* of toxic waste.
>
> **THE RIGHT NOUN:** Our task was to remove *mounds* of toxic waste.

Of course, you will at times use adverbs and adjectives, and many can add zest and precision to your writing. But if you try doing without them, you will use only the best of both and strengthen your writing.

Two words that should send shivers up your spine are *very* and *really*. Cast aside these crutches for feeble sentences.

> WEAK: The program was *very* good.
>
> STRONGER AND MORE PRECISE: The program was *informative* and *inspiring*.
>
> WEAK: We expect trainees to *really* know this material.
>
> STRONGER AND MORE PRECISE: We expect trainees to *master* this material.

Just Try It

Rewrite these sentences, deleting as many adverbs and adjectives as possible, but keep the meaning intact, or strengthen it.

1. They reacted hesitantly before deciding on the promotion.

2. The recruitment brochure gave high visibility to employee benefits.

3. Most retail companies today are having a very hard time attracting and keeping good workers in the stores.

In Your Writing Today ...
☞ **When you can, avoid adverbs and adjectives.**

59. How to Interview

"Interview? I'm not a journalist," you must be thinking. True, but do you ever find it necessary to question people for information and use it in a memo, letter, proposal, or report? To get that information you discuss the subject with, or "interview," the person with the information.

A good interview begins long before you speak to the interviewee. You prepare yourself by learning all you can about the subject on your own. This background enables you to grasp the subject faster, ask better questions, and assimilate information.

Think about the subject and frame some key questions that need to be answered. These questions will depend in part on the end product, the document you plan to write. For instance, for a market research report, you'll have questions about types of products, numbers of installations, success stories, plans for new products.

Limit your questions to five or six, and phrase them open-endedly. That is, don't ask questions that will elicit one-word answers:

"Is the objective of the project image-building or profitability improvement?"

Instead, ask questions that prompt the interviewee to answer in depth:

"What are the major objectives of the project?"

Even when you know certain key information, don't assume too much. People may surprise you, or at least give you more detail, which makes your writing job easier.

In the interview, take the lead from the interviewee. Some will "take over" immediately and start telling you exactly what they want. In that case, just follow, and ask questions only for clarification. At the end of the "monologue," review your list of questions and ask any that haven't been addressed. Other people may wait for your lead. In that case, begin the interview with your first question.

As your interviewee speaks, take notes in some form of shorthand. Don't feel uncomfortable about this. Your interviewee will respect your desire to record his or her thoughts accurately.

When you both feel the interview is over, review quickly what you have. This review ensures your getting all the details right and prevents misunderstandings. As you close, stay alert. People sometimes remember small but important details after the interview is over.

Just Try It

Prepare a list of questions for an interview with:

1. A potential client who needs services your company provides.

2. A superior seeking a two-year plan for your job or department.

In Your Writing Today ...
☞ **Answer the right questions.**

60. Subordination: Not All Ideas Are Created Equal

Inequality is a fact of life in business. The organization chart clearly shows that some employees are subordinate to others. The same is true with ideas. Some are less important than others and should not be allowed equal prominence in your writing.

Just as subordinates serve a useful function in business, subordinate ideas are necessary in writing. The problem arises when writers fail to distinguish between important and subordinate ideas, letting the reader focus attention on subordinate ideas rather than the main ideas.

> NO SUBORDINATION: The growing trend is to compensate top management according to company and individual performance. Compensation in many companies is still based on criteria such as market rates, personality, or seniority.

What we have here are two ideas going in different directions, and without a clue as to which is the main idea. The second sentence in the example is *not* critical to the main idea of the paragraph. Yet because it is written *as a sentence*, it seems to be just as important as the other. To lessen the stature of the second idea, express it as *part of* the other sentence to subordinate it.

> SUBORDINATION: Though compensation in many companies is based on criteria such as market rates, personality, or seniority, the growing trend is to compensate top management according to company and individual performance.

Note how the idea of "the growing trend" becomes the top banana and the idea of "compensation in many companies" plays only a supporting role.

Challenge the importance of each idea. If you decide some ideas deserve subordinate status, work them into the other sentences. Usually you can do that by putting the subordinate idea in a subordinate clause (how clever!). Make the clause subordinate by introducing it with a word such as *if, when, although, because,* as in these examples:

> *When* the price of oil fell, tremors reverberated throughout our industry.

> Management had no contingency plans to deal with the collapse, *although* signs of the fall had been imminent.

Just Try It

Rewrite the following paragraph, subordinating less important ideas.

> The firm as a whole is doing quite well this year. Our office, however, is not doing as well. We posted good numbers last year, but slipped this year. People have offered a number of explanations for the decline. The most valid of these is our loss of key personnel. We did recruit some excellent replacements, but you can't escape a letdown when you've lost top-notch, experienced people.

In Your Writing Today ...
☞ **Highlight key ideas by subordinating other ideas.**

✔ Writing the PD (Practice Development) Article

If you are serious about your career, you probably read trade and professional journals regularly. Have you ever noticed who writes the articles in these publications? A good portion of them are written by people like you, people with an interest in furthering their careers and developing their businesses. They are good at what they do, but not necessarily renowned experts in the field—until they get published, at least. It's amazing how much a published article can do for an individual, within the firm and profession, and with customers.

Regardless of your field, there is probably at least one publication your customers and target accounts read regularly. As they read, they look for new ideas and solutions. Even when they aren't looking for anything specific, they often hit upon a suggestion they can use to improve their operations. When they find that helpful idea in an article with your name on it, you might very well get a call.

How to Write the Article

First, choose a topic. The best source for article ideas is your work—the jobs you've done, the problems you've solved. Of those, which is the most likely to show up as a problem in other companies? That issue is a prime candidate for an article.

Once you've decided on a topic, outline your article. Create subheadings, such as Needs, Solutions, Benefits, Costs, and Prices, and write a short statement of what you will say in the article.

Third, contact the publication your target audience reads. By letter or phone, "query" the editor. That is, tell the editor the kind of article you want to write, briefly what it contains, your qualifications for writing it, and, most important, why the readers of that publication would benefit from it. The editor responds with a yes, no, or maybe.

With the go-ahead, you write the article. The key is finding the readers' "hot buttons." If you just try to hype your service or product, your article probably will not be accepted. So slant your ideas toward the needs of the readers. Focus on their problems and mention your service, equipment, or experience only in terms of solving those problems. Write an *informational* article, not a marketing piece.

After hooking the readers at the beginning with a description of their problems, work through the material in your outline. Be brief, but give enough details and examples for readers to understand your ideas in a *concrete situation*. If you have to present theory, illustrate it with real-world examples. Write in a direct, simple style, and include charts or other graphics that help convey your message. Then proofread carefully.

In Your Writing Today ...
☞ **Think about writing for publication.**

Recommended Answers and Revisions for Exercises

1
Hook the Reader

Rewrite these sentences to hook the reader through self-interest.

> 1. We at ACB have developed new products and services to deal with the technological changes assaulting the widget industry.

Revision: The widget industry, as you well know, has been assaulted recently by a number of technological changes. Among the most challenging of these are . . .

> 2. *Office News* needs information on the happenings and hobbies of people in our office. Help us produce a great newsletter by sending a few paragraphs of your activities.

Revision: What you do, in and out of the office, benefits you and the company. Yet how many people in the firm know the real you? How many know of the real contribution you are making? One way to raise your visibility and your "stock" is through *Office News*.

Write the opening to a memo explaining to a superior why you should be given greater responsibility in the department.

> The Wallens contract has no doubt increased the demands on your time. And Louisa Jackson's pending retire-

ment will put an even greater strain on you to administer the office. Therefore, I suggest that while Louisa is still here, I begin to assume some of her administrative duties. This move will ensure a smooth transition and relieve you of a burden you obviously do not need at this time.

2
For Clarity, Write Short Sentences
Improve the clarity of the floods of words in these sentences by shortening and damming each with a period and appropriate word changes.

1. The software plan must include a number of recommended applications (such as spreadsheet and other applications) and must address the development of uniform standards so that the information systems department does not have to support multiple technologies for performing the same activity.

Revision: The software plan must include a number of recommended applications, including spreadsheets. Standards must be uniform so that the information systems department does not have to support multiple technologies for performing the same activity.

2. One popular approach to software selection is to select one or two packages for which training and other user assistance will be provided and let users know that if they want to implement other packages for the application, they do so knowing that they will not be able to call on internal resources for support.

Revision: One popular approach to software selection is to select one or two packages that provide training and other user assistance. Let users know that if they want to implement other packages for the application, they will not be able to call on internal resources for support.

3
Avoid Clichés

Rewrite these sentences, replacing the clichés. Your original expressions need not be as picturesque as the clichés, but they should convey the same ideas effectively.

1. Good managers avoid sounding like Monday-morning quarterbacks.
2. The plan seemed workable, but the fly in the ointment was the cost.
3. We can have a ballpark figure for you in a week.
4. Though we may still be wet behind the ears in this business, we recognize a pig-in-a-poke offer when we see one.

Suggested Revisions:

1. Good managers do not challenge decisions with after-the-fact information.
2. The plan sounded workable, but cost was the only problem.
3. We can have an estimate for you in a week.
4. Though we are new to this business, we recognize a pig-in-a-poke when we see one.

4
Get the Most from Each Idea

The main idea of the following paragraph is in the first sentence. Delete information from the other sentences that does not relate directly to the paragraph's main idea. (Deletions are in italics.)

> Leadership changes aided the collapse. *Communication throughout the company was also a problem.* The founder's brother-in-law, Harvey Nepotamkin, served as president from 1962 to 1968. *These men had never really liked each other but were tied together through the wife/sister, which neither really liked either.* Nepotamkin's successor was Mabel Frump. When the ex-

pansion program ran into trouble, Frump was replaced by Sidney Stylish at the end of 1971. *Inventory problems were quite evident during these years and management should have recognized the credit disaster developing.* As sales and profits worsened, company veteran R. J. Fullhouse was summoned from retirement to try to turn things around.

Which ideas should be main ideas in paragraphs of their own?

Communication ... was a problem.

Inventory problems ... were quite evident.

Management should have recognized the credit disaster developing.

5
Make Bulleted Items Parallel

Recast one or more bullets in each sentence below to make all bullets parallel. Also make them grammatically correct.

1. Duties of the department include:
- maintaining files on all clients
- documentation of expenditures
- to report irregularities in office practices.

Possible revisions:
- maintaining, documenting, reporting
- *Or* to maintain, to document, to report
- *Or* maintenance of, documentation of, reports of

2. The report will provide:
- an analysis of each key issue
- profiles of the market and consumers
- discuss the options open to the company

Most efficient revision:
- a discussion of the options open to the company

6
Highlight Ideas with Phrasing Patterns

Use *repetition* in a paragraph to stress the point that *detailed planning* led to these positive results.

- The building program is ahead of schedule
- We have controlled costs
- Construction is expected to be completed within budget

Revision: Detailed planning has kept the building program ahead of schedule. Detailed planning has enabled us to control costs. Detailed planning will complete construction within budget.

Use *antithesis* in a sentence to show the difference between two approaches to a job you do.

Example: Rather than being a person who thinks things up, I try to be a person who gets things done.

7
Use Comma Sense

Insert or delete commas to make these sentences clearer.

1. Managers, who straddle the fence, eventually find it uncomfortable.
2. When working the machine produces excellent copies.
3. If all items match the purchase order system schedules payment.

Correct:

1. Managers who straddle the fence eventually find it uncomfortable.
2. When working, the machine produces excellent copies.
3. If all items match, the purchase-order system schedules payment.

8
Spelling Does So Count!
Proofread this paragraph for *ten* misspelled words. Underline them and write the correct spellings below.

> The inventory problem was *intensifyed* by *inefficeint managment*. *Inadequatly*-trained managers were *oparating thier* stores almost *independantly*. Buyers had to rely on *inacurate* sales figures which *resultd* in *to* many under- or over-stocked conditions.

Correct spelling.

intensified	their
inefficient	independently
management	inaccurate
inadequately	resulted
operating	too

9
Anchor the Floating "This"
Anchor the floating *this* in these sentences.

1. We have devised a unique approach to deal with the issue of employees retention. This **approach** is best described in stages.
2. In recent months the personnel department has emphasized attitude over skills in hiring entry-level staff. This **change** has already shown some positive effects.

10
Repeat Key Ideas, or Once Is Not Enough
Write a closing sentence for this paragraph that repeats (paraphrases) its main idea.

> Despite the costs, turnover does at times have positive aspects. For example, turnover provides the opportunity to replace below-average performers with above-average performers. When such a switch is made, productivity can increase greatly, more than making up

for the costs of terminating, recruiting, and training. **Turnover, then, should not always be seen as a totally negative phenomenon.**

11
Letters That Remove the Chill from Cold Calls

Rewrite the beginning of this sales letter.

As pioneers in the widget industry, we at WOW Widgets have always tried to meet the needs of our customers. We continue this practice with a new service line for inventory control called WOW-I.

Improved:

Inventory control in the widget industry has always been critical to success. It dictates production runs and distribution schedules. Yet it has been time-consuming and costly.

12
Beware of Series of Qualifiers, or Don't String the Reader Along

Rewrite these sentences, putting the qualifiers after the key word.

1. Segregating duties will help ensure the proper safeguarding of cash receipts and insurance premium collections.

2. A committee has been formed to examine long-term strategy and short-term objectives costs and benefits.

CLEARER:
1. Segregating duties will help ensure the proper safeguarding of **collections of** cash receipts and insurance premiums.

2. A committee has been formed to examine **costs and benefits of** long-term strategy and short-term objectives.

13
Keep the Verb Close to the Subject

These sentences suffer from subject-verb separation. Rewrite them to make them tighter and clearer. Restructure the sentence if necessary.

1. A review of relevant statistics, such as orders placed, percentage of out-of-stocks, numbers of damaged shipments, frequency and length of delays, and level of bad-debt experience with each supplier, helped us assess our supplier relationships.

2. One common practice that companies opening new branches engage in to ease the burden of large pre-opening costs is to amortize pre-opening expenses over a period of six months to two years.

TIGHTER:
1. A review of relevant statistics helped us assess our supplier relationships. This review included items such as orders placed, percentage of out-of-stocks, numbers of damaged shipments, frequency and length of delays, and level of bad-debt experience with each supplier.

2. To ease the burden of large pre-opening costs, it is common for companies opening new branches to amortize pre-opening expenses over a period of six months to two years.

14
Details, Definitions, and Examples Make Abstract Ideas Concrete

Rewrite this passage, clarifying ideas with details, definitions, and examples.

> The widget industry is becoming more competitive. (*Details:* **There are more companies in the business and price wars are a standard practice**. Recent mergers and acquisitions have created a protectionist mentality among management. (*Definition:* **Managers**

base decisions not on what is good for the company but on the company's attractiveness for takeover.) Established companies have taken to some very questionable practices. (*Example:* **For example, WOW widgets, in the business 50 years, has suddenly gone deeply into debt by diversifying into the underwear business.**)

15
Omit Unnecessary Words

Omit the unnecessary words from these sentences.

1. Their annual contribution has been $1,000.
2. They developed a comprehensive job description.
3. Their forecasts of sales trends are overly optimistic.
4. The program helps managers prepare the plan and monitor its success.

16
Avoid Wordy Expressions

1. What verbs can replace these wordy noun/verb expressions?

make provisions for	provide
provide documentation	document
undertake completion	complete

2. Provide a word or shorter phrase for each of the following:

subsequent to	after
due to the fact that	because
with the exception of	except

17
The Topic Sentence: A Guidepost for Readers

Tell what's wrong with these topic sentences.

1. We added seventeen people to the office staff in the last year. (NOT IN GENERAL TERMS)
2. Our company has a bottom-up approach to planning and a zero-based budgeting philosophy. (TOO MANY IDEAS)

Rewrite the second sentence above as a topic sentence and build a paragraph around it.

> *Our company has a bottom-up approach to planning.* There are regular meetings between staff and upper management, at which ideas about problems and needs are exchanged. To develop annual budgets, management asks departments for two lists: one for immediate needs to allow for continued performance of tasks at current levels. The second is a "wish" list that itemizes what would be needed to change procedures and improve performance. Management considers both and discusses them with the staff before making final decisions.

18
How to Build an Outline: The Beams and Girders of a Report

Complete the outline started in Lesson 18. Consider the additional major points to be "current status" and "plans for immediate future."

The outline might look like this:

    An Assessment of the Company
I. Company History
   A. Early Formation
   B. Expansion
      1. Territorial
      2. Products and services
         a) consumers
         b) businesses

    II. Current Status
        A. Sales
            1. Services
            2. Products
                a) standard
                b) new
        B. Market Share
    III. Plans for the Immediate Future
        A. Increase Market Share
            1. New Units
            2. Additional Sales Force
        B. Investigate Diversification into New Businesses

19
The Hyphen: You've Come a Long Way, Baby
Insert, delete, or reposition hyphens as needed in these sentences.

1. The company is now planning to reinstate the eight cylinder (**eight-cylinder**) engine.
2. Twenty four (**twenty-four**) staff people attended the three day (**three-day**) meeting that was int-ended (**intend-ed**) to unveil the well-kept secret of the home-shopping (**home shopping**) subsidiary.
3. In this labor-intensive environment, lay-offs, (**layoffs**) are common.

20
Transitions: Bridges to Understanding
1. Use a transitional word or phrase to link these two sentences:
    Office romances have always been thought to affect productivity adversely. Some psychologists, *HOWEVER*, claim people in love with co-workers have better morale in the office and better attendance records.

2. Repeat a key word or idea to link these two sentences: Though the content of the brochure is acceptable to the committee, there are some serious disagreements about the format. **To iron out these differences of opinion or to address these disagreements**, a meeting will be scheduled for Monday at nine.

21
The Analogy: Explaining What It Is by Telling What It's Like

1. Write a single sentence that uses analogy to explain the idea that a direct approach is often the best approach.

   I suggest you tell him exactly what's on your mind. You get a heartier welcome if you enter through the front door.

2. Write a paragraph built around an analogy. The main idea to be developed in the paragraph is your concept of a good manager.

   A good manager is much like a good gardener. He must look at the whole plot of ground, not just one parcel of it. He must know when to encourage plants along with food and water supplements and when to just give them time to follow nature's course. He must know when to weed and prune away the dead and dying before they choke off the healthy. And he must recognize when the fruit is ripe and ready for picking.

22
Good Beginnings

Write the beginning of a memo from the office manager to the staff announcing a change in an administrative procedure. Indicate which of the four objectives of a good beginning you have reached.

   So that you can be reimbursed for expenses more quickly, we have instituted a new procedure for reporting expenses. The reasons for the changes and the steps to follow in the procedure are explained in this memo. Sample forms are attached. (This beginning tries to generate interest and tell what's covered.)

23
The Middle: Delivering What's Promised
Review a document of three or more pages. Determine if: (1) the middle delivers what the beginning promises; (2) the most important elements are given the most space; and (3) the order makes reading easy.
Answers will vary here.

24
Writing the Ending, or Leave 'em Laughing
1. Write the ending to a memo on office courtesy that makes a call for action.

    Everyone realizes that under the stress of a busy season we all get testy from time to time. But I ask you to be patient, count to ten, and remember that courtesy is contagious.

2. Write the ending to a letter of application that conveys your enthusiasm for working for the company.

    Yours is just the kind of dynamic organization I have been seeking and in which I can truly use my skills and grow. Therefore, I ask you to consider my application seriously. I anxiously await your call.

3. Write the ending to a report on turnover in your organization that states the significance of your findings.

    The numbers, then, clearly show that our turnover problem is largely a recruiting problem. We lose people after only a short time because we are not hiring the right people in the first place.

25
Keep It Simple: Conversational Language

Write a simpler, conversational word or phrase for the following:

> rectify: **correct**
> aforesaid: **mentioned above**
> strategize: **plan strategy**
> headcount reductions: **layoffs**
> effect a change: **change**
> regarding yours of: **in regard to your letter of**
> endeavor (verb): **try**
> ameliorate: **improve**
> herewith are: **here are**

26
Add Vitality with the Active Voice

Convert these sentences into the active voice *where preferable.*

1. The case will be tried by one of our senior partners.
   One of our senior partners will try the case.

2. More than a thousand customers were issued inaccurate statements.
   Acceptable as is; doer is less important than receiver of the action.

3. A decision has not yet been reached by management.
   Management has not yet reached a decision.

4. Inactive files should be purged from the system daily.
   The Head Filer should purge the system of inactive files daily.

27
Action Verbs Add Snap

Enliven these sentences by replacing weak verbs with action verbs.

1. We provide assistance to companies without their own PR departments.

161

      We **serve** companies without their own PR departments.
2. The opponents discussed the issues for hours without any progress.
    The opponents **debated** the issues for hours without any progress.
3. Computer malfunction was the cause of the erroneous billings.
    Computer malfunction **triggered** the erroneous billings.

28
Make References Clear

Rewrite these sentences to make the references clear.

1. When Harry presented the plan to Don he was happy.
   Don was happy when Harry presented the plan to him.
   Or: Harry was happy when he presented the plan to Don.
2. Edit the introduction to the report and add some graphics to it.
   Edit the introduction and add some graphics to the report.
   Or: Edit and add some graphics to the introduction to the report.

29
Who, Which, *and* That

Correct the misuse of pronouns and commas in these sentences.

1. The investor (**who**) bought the company had once been its mail clerk.
2. Wall Street analysts (**who**) had predicted the takeover were ecstatic. (No commas)
3. Total costs, which can only be estimated, approached $1 billion. (Commas)

162

4. An effect (**that**) no one foresaw was the wholesale firing of management.

30
Writing Numbers: Words or Figures?
Assuming the following sentences come from different documents and each contains only a few numbers, change those numbers where necessary to conform to the guidelines given above.
   1. Last week 6-month (**six-month**) T-bills sold at an average rate of 7 (**seven**) percent.
   2. 120 days after the company had been bought, it was sold again.
      The company was resold 120 days after it had been bought. *Or:* **Just 120 days after it had been bought, the company was sold.**
   3. The average family no longer has 2.5 children; the average child now has 2.5 parents. (*Correct*)

31
Persuasion Starts with the Reader
   1. How would you try to shape your message if your reader were analytical?
      If she is analytical, give her plenty of facts and details so she can convince herself with your data that you are right.
   2. If she were *authoritarian?*
      If she is authoritarian, mention titles, degrees, experience, and quote experts and other authorities whose words support your thinking.
   3. If the management team repeatedly mentioned words like *efficiency, productivity, streamline?*
      Be sure to highlight things like reduced staffing needs, increased output, lower costs.

32
Positioning Ideas in Sentences

Rewrite these sentences as directed.

1. As last year, increased productivity is the goal for the department this year. (Consider **productivity** the most important idea; **goal** is the second most important idea.)

    Our goal for the department this year is the same as it was last year—increased productivity.

2. After auditing the department, we find that none of the recommendations we made last year has been implemented and that for the second straight year the department is grossly inefficient. (Most important: **recommendations not implemented**; second most important: for the **second straight year.**

    For the second straight year our audit finds the department grossly inefficient, and we find that none of our recommendations of last year has been implemented.

33
Imagery Makes Readers See the Point

Rewrite these sentences, using a visual image to express an idea.

1. For weeks the department had been trying unsuccessfully to process an unusually high number of requests.

    For weeks the department was drowning under wave after wave of requests.

2. The marketing plan includes a series of different ads to reach the public through various media in a concentrated period of time.

    The marketing plan calls for a multi-media blitz.

34
You Said It! Using Direct Quotations
1. Reread some of the reports, letters, and memos you have written Try to insert an appropriate quote from the intended reader, an expert, literature, or any other source.

2. Thumb through a book of quotations and note those that might have been useful in documents you have written. Jot some down for future use.

Answers for both questions will vary with the writer.

35
... And Everything in Its Place (Misplaced Modifiers and Dangling Participles)
Rewrite these sentences to correct misplaced modifiers and dangling participles.

1. I can count the number of times I've sat before my personal computer on one hand.

I can count **on one hand** the number of times I've sat before my personal computer.

2. We assist clients, acting as facilitators, in negotiating contracts.

Acting as facilitators, we assist clients in negotiating contracts.

3. Realizing a bold move was called for, the new product was announced three weeks early. (Hint: Who is realizing?)

Realizing a bold move was called for, the company announced the new product three weeks early.

36
Variety Is the Spice of Sentence Structure
Rewrite this paragraph, creating variety in sentence pace, length, and word order.

165

Management holds planning meetings each month. All department heads attend these meetings. Each makes a report on his or her department. Management presents company-wide figures. Management then sets company and department goals for the next period.

ONE POSSIBLE ALTERNATIVE: Each month, management holds a planning meeting. All department heads attend these meetings, reporting on their departments. Then management presents company-wide figures and sets goals for the next period for the company and departments.

Quicken the pace of this sentence to match the idea behind it:

The new procedure will reduce delay in our responding to inquiries.

The new procedure will generate quick response to inquiries.

37
Distant Relatives: The Colon and Semicolon

Insert or interchange colons and semicolons where appropriate.

1. Our press release was picked up by the following newspapers;(:) *Graceland News, Valley Shopper,* and *The Sentinel.*
2. The event included speakers from companies such as Coca Cola, IBM, and General Motors,(;) journalists for television, radio, and the general and business press,(;) and attendees from companies throughout the world.

38
Unconventional Techniques Perk Up Business Correspondence

1. Write a rhetorical question introducing reasons for a

new procedure you have described earlier in the document.

   EXAMPLE: **Why the change in procedure?**

2. Inject alliteration into this sentence below. You may change words. The two glass buildings stood next to each other

   EXAMPLE: The two gla__ss__ __s__tructure__s__ __s__tood shoulder to shoulder.

3. Inject onomatopoeia into this sentence.
   After two bad quarters, the company came back strong in the third.

   EXAMPLE: After two bad quarters, the company **rebounded** in the third.

39
Dashes and Parentheses: The Interruptors
Punctuate these sentences with parentheses or dashes. Suggested versions are:

1. To prevent last-minute deadline crunches—which are happening on a regular basis—we have installed a new schedule for the newsletter.

2. The candidates (six in all) will be in the office at nine o'clock on Saturday for a full round of interviews.

3. Traffic on Route 197 has increased substantially in the last year. (We have been tracking it with funds from a federal grant.) Residents see that increase as a torture, but businesses see it as a blessing.

40
Avoiding Sexist Language
Rewrite this paragraph to avoid sexist language.

   Normally a project of this sort requires at least 180 manhours (**workhours**). We interview everyone from the salesmen (**sales representatives**) on the road to the girls (**switchboard operators**). We ask each person (**all people**) in the department to describe his (**their**) job(**s**) and ask if he (**they**) has (**have**) any suggestions for improving his (**their**) performance.

41
Building a Working Vocabulary
1. Write a sentence for each of these words: *propel, devoid, cites, frozen, impetus,* and *emboldened.*

    Helping to *propel* this growth is an influx of immigrants.

    The transaction was *devoid* of contracts and lawyers.

    Management *cites* the factor of declining sales.

    Salaries were *frozen* for the second year.

    Unpleasant conditions can be the *impetus* to move on.

    Emboldened by desperation, they achieved a great victory.

2. Grade yourself on these. Have an extra nice lunch if you do well.

42
Developing Ideas
Which methods could you use to develop each topic-sentence idea?

1. The project can be accomplished in three phases. (*Classification*)

2. I see the domino theory at work in this situation. (*Definition, explanation, illustration, causal relationships*)

3. Management has installed new procedures to combat pilferage. (*illustration, causal relationships, description*)

168

4. We have made a strong commitment to the cities we operate in. (*Definition, illustration*)

5. One philosophical concept permeates our entire organization. (*Definition, explanation, illustration*)

Develop one of the ideas above using the methods you identified.

44
Complete Agreement

Correct *errors* of agreement in this paragraph.

> The meeting of company stockholders were (**was**) held on Monday. The Chairman and the management team was (**were**) present to answer questions, most of which revolved around rumors of plant closings. The Chairman or an appropriate designee were (**was**) evasive in discussing the issue. Management apparently will not tip their (**its**) hand prematurely.

45
Fragments

Rewrite these sentences, correcting the sentence fragments.

1. The direct mail piece will focus on a case study. A project we completed successfully for a client.

   The direct-mail piece will focus on a case study, a project we completed successfully for a client.

2. When you asked us to propose on this subject, back in September, to research the market to determine the feasibility of your introducing your X224 to the industry.

   When you asked us to propose on this project back in September, to research the market to determine the feasibility of your introducing your X224 to the industry, **you said nothing of a marketing plan.**

3. Your letter called for a quick, summary-level response. We sent it.
(*Correct*)

46
Drawing Comparisons

Write a sentence using a metaphor or simile to help explain each of the following. Identifiy the method of comparison used in each sentence.

1. the department's poor attitude
2. a manager's boring memos

SAMPLE POSSIBILITIES:
1. Each morning the department members approached their work like schoolchildren called in from recess. (*Simile*)
2. His memos, with all the bulk and excitement of the telephone book, appeared almost every day. (*Metaphor*)

48
Common Errors in Word Usage

Correct all errors in word usage.

What we imply (**infer**) from your letter is that its (**it's**) our responsibility to head up (**head**) the Committee for Restitution. The Committee will be composed (**comprised**) of less (**fewer**) than six people. Its objective will be to impact (**affect**) the attitudes of all citizens we meet up with (**meet**) during the year.

50
Abbreviations: The Long and Short of It

Make the appropriate changes in the use of abbreviations in this paragraph. Where the abbreviation is inappropriate, remove it.

We received a letter from Polly Esther, Ph.D. (**Ph.D.,**) a mgr. (**manager**) working in the Consumer Relations Dept. (**department**) for the E.P.A. (**EPA**). Her letter alludes to our procedures for dumping waste, etc. (**removing it, notifying authorities, and reporting irregularities**). Ms (**Ms.**) Esther warns that we are in violation of several federal regulations. Some of these regulations, i.e., (**e.g.,**) those on toxic waste, carry severe penalties.

51
Run-Onnnnnnnnnnns
Correct these run-ons.

1. The job entails providing many documents simultaneously (**conjunction and comma, semicolon, or period**) it requires someone with good organizational skills.
2. The individual in this position must realize it is a service function, (**conjunction and comma, semicolon, or period**) he or she must be receptive to requests for help.
3. Good service is our business (**conjunction and comma, semicolon, or period**) business is good.

52
Discriminate and Delete
Delete unnecessary words and ideas from this passage.

Working a convention to develop leads and business takes experience and skill. Training in correct behavior can provide some of the skill. For one thing, one must arrange the booth attractively to draw conventioneers to it. A well-laid-out booth will (**go further and**—*cut*) pull those conventioneers (**passing by**—*cut*) onto the carpet and into the booth. The sales rep (**working the booth**—*cut*) then must know how long to let the person browse before approaching, and how to offer assistance. The wrong approach will **(cut off the visitor's interest and cause**—*cut*) (*insert* **drive**) him or her (**to leave**—*cut*) (*insert* **from**) the booth immediately.

53
More Troublesome Words
Correct the errors in this passage.

The meeting was held to orientate (**orient**) staff of the acquired firm. New management explained that as people of principal (**principle**), the negotiating parties had reached only a verbal (**an oral**) agreement and that irregardless (**regardless**) of rumors, over (**more than**) 75 percent of the staff would be retained. Any employee who's (**whose**) job was eliminated would receive a very (*cut* **very**) unique form of severance.

54
Proofreading Pays Off
Proofread this passage and correct all errors.

Perhaps we're (**we've**) not had many sales with this service line because there is not (**no**) need for it. Are we trying to sell something people don't need or want.(?) Are we pushing the technlogy (**technology**) without showing what business problems the technology solves? I feel—as I have all along (—) that this (/) service is of value to clients and will be profitable for us(.)

55
Writing Directions
Rewrite the directions that follow, using the suggestions in the lesson.

To have the mailing list updated, these steps should be followed:

1. The current mailing list should be reviewed for errors.
2. Names no longer relevant to our purposes should be deleted.
3. The edited lists should be returned to me.
4. Modifications on names, addresses, titles, etc., should be made.
5. Names not on the list that you want on should be added.

   So that we can make all of our mailings productive, (give a reason for doing this) we have to update our lists. To do so, please follow these steps(:) (*Active, not passive, voice*)

   1. **Review the current mailing list for errors.** (*Active*)
   2. **Delete names that are no longer relevant to our purposes.** (*Active*)
   3. **Add names that you** (*Make it personal*) **want on the list but are not on it.** (*Active*)
   4. **Modify names, addresses, titles, etc.** (*Active*)
   5. **Return the edited lists to me.** (*Active, logical sequence*)

56

More on Sexist Language

Rewrite the following paragraph, removing the sexist language.

Normally a project of this sort requires at least 180 man-hours (*workhours*). Our project director, a woman (*delete*) management specialist, interviews everyone from the salesmen (*sales representatives*) on the road to the girls (*operators*) at the switchboard. To comply with EEO regulations, we ask the ladies (*women*) if they feel they are treated fairly. We ask the group to identify a spokesman (*spokesperson*) who will report on department-wide complaints.

57
Setting the Right Tone

Rewrite these statements to achieve the tone indicated.

1. As per your letter of 7/2, the following decisions have been made. (*Rewrite with more informal tone.*)

   We have made some decisions in the matters you pointed out in your letter of 7/2.

2. Before the sheriff comes to throw us out and bar the door, let's see what we can salvage. (*Rewrite in less whimsical tone.*)

   Despite the seriousness of our situation, we should meet to discuss ways of minimizing our losses.

Write a short memo announcing to your department that the company is being sold. Strike a serious yet non-threatening tone. Indicate that people may lose jobs, but try to assuage people's fears.

Answers may vary. Here is one approach.

> As you may have heard, the company is being sold to Janice & Co. The sale and exchange will take place on September 1; until then, and perhaps for a good time after, we will operate as we have in the past. While some of us may be replaced by Janice people, there is a good possibility many of us will be asked to stay on. It cannot hurt to start looking, but there is no need to panic. I have scheduled a meeting for Tuesday at 10 A.M. to give you more details and to discuss your concerns.

58
Advice on Adverbs and Adjectives

Rewrite these sentences, deleting as many adverbs and adjectives as possible, but keep the meaning intact, or strengthen it.

1. They reacted hesitantly (**hesitated**) before deciding on the promotion.
2. The recruitment brochure gave high visibility to (**featured**) employee benefits.
3. Most retail companies today are having a hard time attracting and keeping (**struggling to attract and keep**) good (*dependable*) workers in the stores.

59
How to Interview
Prepare a list of questions for an interview with:

1. A potential client who needs services your company provides.
   Answers will vary but here are some possibilities:

   What is your objective?

   How do you think it can be reached?

   How do you think we can help?

   When would you expect this to be done?

2. A superior seeking a two-year plan for your job or department.
   Answers will vary but here are some possibilities:

   What assumptions can we make about staff, in terms of numbers and individuals?

   What can we expect in terms of budget?

   What company-wide goals should we know about that will affect the department?

60
Subordination: Not All Ideas Are Created Equal
Rewrite the following paragraph, subordinating less important ideas.

The firm as a whole is doing quite well this year. Our office, however, is not doing as well. We posted good numbers last year, but slipped this year. People have offered a number of explanations for the decline. The most valid of these is our loss of key personnel. We did recruit some excellent replacements, but you can't escape a letdown when you've lost top-notch, experienced people.

Answers may vary. Here is one possibility.

Though the firm as a whole is doing quite well this year, our office isn't. Unlike last year, when we posted very good numbers, we slipped this year. Among the number of explanations offered for the decline, the most valid explanation is the loss of key personnel. Despite our recruiting some excellent replacements, you can't escape a letdown when you've lost top-notch, experienced people.

Index

Abbreviations
  in business writing, 120–121
  numbers with, 72
  punctuation with, 120
Abstractions, 34–35
Action verbs, 66–67
Active voice
  in business writing, 64–65
  in directions, 132–133
Addresses, 72
Addressing readers, 51
Adjectives
  compound, 46–47
  with participles, 47
  in sentences, 140–141
Adverbs, 140–141
Advertising techniques, 93
Affect, 116
Affixes, 46–47
Agreement, subject-verb, 106–107
Alliteration, 92–93
Analogies, 52–53
And, 50
Antithesis, 16–17
Appearance, 15
Articles, informational, 146–147
Audiences. *See* Readers
Average, 128

Bastardized words, 61
Be, 66
Because of, 116
Beginnings. *See* Openings
Body. *See* Middles
Boilerplates, 110–111
Brainstorming, 38–39. *See also* Ideas
Bulleted items
  in business writing, 12–13
  punctuation with, 90

Business correspondence. *See* **Business writing**
Business writing. *See also* Reports; specific techniques of
  abbreviations in, 120–121
  action verbs in, 66–67
  active voice in, 64–65
  amount of, on job, 2
  analogies in, 52–53
  appearance of, 15
  boilerplates in, 110–111
  brainstorming in, 38–39
  bullet-point lists in, 12–13
  clichés in, 8–9
  comparisons in, 112–113
  correctness in, 2
  criteria for, 14–15
  deletions in, 126–127
  directions, 132–133
  directness in, 2
  endings in, 58–59
  form letters, 110–111
  idea development in, 102–103
  imagery in, 80–81
  information gathering for, 114–115
  interviews, 142–143
  jargon in, 118–119
  middles in, 56–57
  numbers in, 72–73
  openings in, 54–55
  outlines for, 44–45, 115
  PD article, 146–147
  personal pronouns in, 50–51
  persuasion in, 76–77
  persuasiveness in, 2
  process of, 26–27
  proofreading, 130–131
  proposals, 122–123
  punctuation in, 86–87

Business writing, *continued*
  qualifiers in, 30–31
  quotations in, direct, 82–83
  readability test for, 104–105
  rewriting in, 74–75
  sales letters, 28–29
  sexist language in, 96–97, 136–137
  this in, 22–23
  time necessary for, 62–63
  tone in, 138–139
  topic sentences in, 42–43
  troublesome words in, 128–129
  unconventional techniques in, 92–93
  on word processors, 134–135
  word usage errors in, 116–117
But, 50

Cataloging information, 114–115
Chronological order, 56–57
Clarity
  commas for, 18–19
  and qualifiers, 30–31
  of references, 68–69
  short sentences for, 6–7
  subject-verb placement for, 32–33
  and S-V-O order, 88–89
  and *this*, 22–23
Clauses
  essential, 70–71
  nonessential, 70–71
  semicolons in, 91
  subordinate, 145
Clichés, 8–9
Cold calls, 28
Collecting information, 114
Collective nouns, 106
Colons, 87, 90
Commas
  for clarity, 18
  and dashes, 94–95
  and run-on sentences, 124–125
  and semicolons, 91
  in sentences, 86
Comparisons, 112–113
Compose, 116
Compound subjects, 106
Compound words, 46–47
Comprehension. *See* Clarity
Comprise, 116
Computers, 134–135
Conclusions. *See* Endings

Conjunctions
  and run-on sentences, 124–125
  sentences beginning with, 50
Connectives, 48
Consistency, 130, 132
Contractions, 98–99
Contrasts, 16–17
Conversational language
  in business writing, 60–61
  and tone, 138
Correctness
  importance of, 2
  modifiers for, 84–85
  participles for, 84–85
  and run-on sentences, 124–125
  subject-verb agreement for, 106–107
Correspondence. *See* Business writing; specific types of

Dangling participles, 84–85
Dashes
  and commas, 94–95
  in sentences, 86–87, 95
Dates, 72
Definitions, 34–35
Deleting ideas, 126–127
Demonstrative pronouns, 22–23
Details, 34–35
Dickens, Charles, 36
Dictionary, 20
Directions, writing, 132–133
Directness
  expressions for, 40–41
  importance of, 2
  wording for, 36–37
Direct quotations, 82–83
Discriminating ideas, 126–127
Drafting stage, 26–27
Due to, 116

e.g., 120
Effect, 116
Endings, 58–59
Essential clauses, 70–71
Etc., 120–121
Examples, 34–35
Exclamation points, 86
Expressions
  grandiloquent, 60–61
  Latin, 120
  parenthetical, 94
  and wordiness, 40–41

Fewer, 117
Figurative language. *See* Imagery
Figures, 72–73
Figures of speech, 138
Five Ws and H, 78–79
Flaunt, 116
Flavor
　active voice, 64–65
　in good business writing, 14
　imagery, 80–81, 112–113
　phrasing patterns, 16–17
　readability, 104–105
　unconventional techniques, 92–93
Flesch, Rudolf, 104
Flout, 116
Fog Index, 104–105
Formal language, 138
Form letters, 110–111
Fragments
　causes of, 108–109
　use of, 99
France, Anatole, 18

Generating stage
　and brainstorming, 38–39
　purpose of, 26
Governmental agencies, 120
Grandiloquent expressions, 60–61

Head, 116
Headings, topic, 114–115
Head up, 116
Hyphens, 46–47

i.e, 120
I
　in business writing, 50–51
　vs. *you*, 4–5
Ideas. *See also* Main ideas
　abstract, 34–35
　actions behind, 80–81
　analogies for simplifying, 52–53
　brainstorming for, 38–39
　bullet-point lists of, 12–13
　concrete, 34–35
　contrasting, 16–17
　deleting, 126–127
　developing, 102–103
　directions, 132–133
　direct quotations, 82–83
　endings, 58–59

　in good business writing, 14
　highlighting, 16–17
　hooking readers, 1–2
　information gathering, 114–115
　interviews, 142–143
　middles, 56–57
　openings, 54–55
　persuasion, 76–77
　placement of, 78–79
　repetition of, 24–25, 36–37
　sales letters, 28–29
　semicolons with list of, 90–91
　stating, 10–11
　subordinating, 144–145
　undeveloped, 102–103
Imagery
　for readers, 80–81
　types of, 112–113
Impact, 116
Imply, 117
Importance, order of, 57
Infer, 117
Infinitives, 98
Information gathering, 114–115
Interruptors. *See* Commas; Dashes; Parentheses
Interviews, 142–143
Irregardless, 128
It's, 117
Its, 117

Jargon, 118–119
Johnson, Samuel, 82

Kipling, Rudyard, 40

Language
　conversational, 60–61, 138
　formal, 138
　jargon, 118–119
　sexist, 96–97, 136–137
Latin expressions, 120
Length, sentence, 88
Less, 117
Letters
　boilerplates for, 110–111
　form, 110–111
　greeting of, 90
　sales, 28–29
Lists, bullet-point, 12–13, 90
Logical order, 56

179

Main ideas. *See also* Topic sentences
  highlighting, 16–17
  repetition of, 24–25
  stating, 10–11
  in topic sentences, 42–43
Mean, 128
Mechanics. *See also* Proofreading; Punctuation
  abbreviations, 120–121
  colons, 90–91
  commas, 18–19
  dashes, 94–95
  in good business writing, 15
  hyphens, 46–47
  numbers, 72–73
  parentheses, 94–95
  proofreading, 130–131
  semicolons, 90–91
  spelling, 20–21
Median, 128
Meet, 116
Meet up, 116
Metaphors, 112–113
Middles, 56–57
Misplaced modifiers
  and clarity of references, 68–69
  correcting, 84–85
Misreading, 18–19
Modifiers. *See* Misplaced modifiers
Money, 73
More than, 128

Names, 120
Nonessential clauses, 70–71
Note cards, 114
Nouns
  and action verbs, 67
  collective, 106
  compound, 46–47
  qualifiers for, 30–31
  strong, 140
  from verbs, 40–41
  weak, 140
Numbers, 72–73

One-sentence paragraphs, 98
Onomatopoeia, 93
Openings
  hooking reader in, 4–5
  strong, writing, 54–55
Oral, 128

Order
  considerations of, 56–57
  of information, 115
  S-V-O, 88–89
  variety of, 88–89
Organization
  in good business writing, 14
  of ideas, 10–11, 24–25
  of information, 115
  outlines, 44–45
  of topic sentences, 42–43
  transitions in, 48–49
Orient, 128
Orientate, 128
Outlines, 44–45, 115
Over, 128

Pace, 88
Paragraphs, 98. *See also* Endings; Middles; Openings
Parallel structure, 12–13
Paraphrasing
  and repetition of ideas, 24–25
  and transitions, 49
Parentheses, 94–95
Parenthetical expressions, 94
Participles
  with adjectives, 47
  dangling, 84–85
Parts of speech. *See* specific types
Passive voice
  in business writing, 64–65
  in directions, 132–133
PD (Practice Development) article, 146–147
Percentages, 72
Periods, 86, 120
Personal pronouns
  in business writing, 50–51
  and hooking reader, 4–5
Persuasion
  importance of, 2
  methods of, 76–77
Phrases
  lengthy, 40–41
  repetition of, 16–17
  semicolons in, 91
  stock, 60–61
Plurals, 106–107
Prefixes, 46–47
Prepositions, 50
Pre-writing, 135

180

Principal, 128
Principle, 128
Process of writing, three-step, 26–27
Pronouns
　demonstrative, 22–23
　misplaced, 68–69
　personal, 4–5, 50–51
　referents, 22, 107
　relative, 70–71
Proofreading
　importance of, 130–131
　for spelling, 20
Proposals, 122–123
Punctuation. *See also* specific marks
　with abbreviations, 120
　with bulleted items, 90
　proofreading for errors of, 130
　purpose of, 86–87
　and run-on sentences, 124–125

Qualifiers, 30–31
Queries, 146–147
Question marks, 86
Questions
　in interviews, 142–143
　rhetorical, 92
Quinn, John C., 76
Quotations, direct, 82–83

Raise, 128
Readability, 104–105
Readers
　addressing, 51
　hooking, 4–5
　imagery for, 80–81
　interest of, generating, 55
　and jargon, 118–119
　losing, 36–37
　persuading, 76–77
　and topic sentences, 42–43
Recognition, order of, 57
Referents
　agreement of, with pronouns, 107
　clarity of, 68–69
　and *this*, 22
Regardless, 128
Relative pronouns, 70–71
Relevance, 128
Relevancy, 128
Repetition
　of ideas, 24–25, 36–37
　of phrases, 16–17

transitions through, 48–49
unnecessary, 36–37
Reports. *See also* Business writing
　information gathering for, 114–115
　jargon in, 118–119
　numbers in, 72–73
　outlines for, 44–45, 115
　time necessary to write, 62–63
Request for Proposal (RFP), 122–123
Revising
　purpose of, 74–75
　stage, 27
　on word processor, 134–135
Rewriting. *See* Revising
Rhetorical questions, 92
Rhythm, 138
Rules
　sentence structure, 50–51, 98–99
　spelling, 20–21
Run-on sentences, 124–125

Sales letters, 28–29
Selecting information, 114–115
Semicolons
　in clauses, 91
　and commas, 91
　in sentences, 87, 90–91
Sentences. *See also* Sentence structure
　adjectives in, 140–141
　adverbs in, 140–141
　analogies in, 52–53
　cliches in, 8–9
　colons in, 90–91
　commas in, 86
　comparisons in, 112–113
　conjunctions beginning, 50
　dangling participles in, 84–85
　dashes in, 86–87, 94–95
　fragments of, 99, 108–109
　idea placement in, 78–79
　incomplete, 108–109
　length of, 88
　misplaced modifiers in, 84–85
　pace of, 88
　parentheses in, 94–95
　phrasing patterns in, 16–17
　prepositions beginning, 50
　punctuation in, 86–87
　qualifiers in, 30–31
　references in, 68–69
　run-on, 124–125
　semicolons in, 87, 90–91

181

Sentences, *continued*
   sexist, 96–97, 136–137
   short, 6–7, 88
   strong, 141
   subject-verb placement in, 32–33
   subordination of ideas in, 144–145
   this in, 22–23
   topic, 42–43, 102–103
   unnecessary words in, 36
   verbs in, 66–67
   voice in, 64–65
   weak, 141
   word order in, 88–89
Sentence structure. *See also* Sentences
   and fragments, 108–109
   in good business writing, 14–15
   idea placement in, 78–79
   modifiers in, 84–85
   parallel, in bulleted items, 12–13
   participles in, 84–85
   and qualifiers, 30–31
   references in, 68–69
   rules, 50–51, 98–99
   and run-ons, 124–125
   short sentences, 6–7, 88
   subject-verb agreement in, 106–107
   and *this*, 22–23
   and tone, 138
   variety in, 88–89
   verb-subject placement in, 32–33
Sexist language, 96–97, 136–137
Short sentences. *See also* Sentences
   for clarity, 6–7
   for pace, 88
Sic, 121
Similes, 112–113
Singulars, 106–107
Spell-check features, 134
Spelling
   aids in, 20
   computer features for checking, 134
   in good business writing, 15
   rules, 20–21
Split infinitives, 98
Stock phrases, 60–61
Structure. *See* Sentence structure;
   Writing structure
Style, 75
Subjects
   agreement of, with verbs, 106–107
   compound, 106
   identifying in opening, 54

   and verbs, 32–33
Subordinate clauses, 145
Subordination, of ideas, 144–145
Substance, 74–75
Suffixes, 46–47
Summarizing, 58, 83
S-V-O order, 88–89

Terms, defining, 34–35
That, 70–71
This, 22–23
Time, writing, 62–63
Times
   abbreviations for, 73
   colons in, 90
Titles, 120
To, 98
Tone, 138–139
Topic headings, 114–115
Topic sentences
   ideas in, 102–103
   main ideas in, 42–43
   for readers, 42–43
Transitions, 48–49

Unconventional techniques, 92–93
Unique, 128
Up, 128
Usage, good, 14–15

Verbal, 128
Verbs
   abstract, 66–67
   action, 66–67
   agreement of, with subjects, 106–107
   concrete, 66–67
   nouns from, 40–41
   plural, 106–107
   singular, 106–107
   strong, 140
   and subjects, 32–33
   weak, 140
Vocabulary, working, 100–101
Voice
   in business writing, 64–65
   in directions, 132–133

Who's, 128
Who, 70–71, 129
Whom, 70–71, 129
Whose, 128

Widows, 126
Wordiness
   expressions causing, 40–41
   unnecessary words causing, 36–37
Wording. *See also* Words
   action verbs, 66–67
   adjectives, 140–141
   adverbs, 140–141
   clichés, 8–9
   comparisons, 112–113
   conversational language, 60–61
   errors in word usage, 116–117
   expressions causing wordiness, 40–41
   in good business writing, 14
   jargon, 118–119
   sexist language, 96–97, 136–137
   that, 70–71
   tone, 138–139
   troublesome words, 128–129
   unnecessary words, 36–37
   vocabulary, working, 100–101
   which, 70–71
   who, 70–71

Word processors, 134–135
Words. *See also* Wording
   abstract, 80
   bastardized, 61
   choice of, 138
   compound, 46–47
   concrete, 80
   different, 49
   division of, 46
   errors in usage of, 116–117
   for numbers, 72–73
   order of, 88–89
   transition, 48–49
   troublesome, 128–129
   unnecessary, 36–37
   vocabulary of, working, 100–101
Writers, identifying, 54
Writing. *See* Business writing
Writing structure, 75
WWWWWH, 78–79

You
   vs. *I*, 4–5
   readers addressed as, 51